MW00804400

Handheld Radio Field Guide

Front Panel Programming (FPP) Instructions
for Handheld Ham Radios (HTs)

Handheld Radio Field Guide

Front Panel Programming (FPP) Instructions
for Handheld Ham Radios (HTs)

By Andrew Cornwall

Listening Bird Press

Handheld Radio Field Guide
Front Panel Programming (FPP) Instructions
   for Handheld Ham Radios (HTs)

Text and images copyright © 2018 by Andrew Cornwall.

Published by Listening Bird Press.

All rights reserved. No part of this work may be reproduced in any form whatsoever without the prior written permission of the copyright owner and the publisher.

Notice of liability: the information in this book is distributed on an "as-is" basis, without warranty. Neither the author nor Listening Bird Press shall have any liability for loss or damage caused or alleged to be caused by use of any information contained within this book.

Trademarked names are used throughout this book. These are used in editorial fashion, with no intent to infringe the trademark. All trademarks are property of their respective owners.

ISBN 978-0-9996609-0-4

First printing

1 2 3 4 5 6 7 8 9 0 – 20 19 18

Contents

Contents

Introduction

There are many different handheld ham radios available today. Your choice will be based on many factors, including features, price, durability and ease of use. Unfortunately, there is no standard for programming a handheld radio—so the steps you'll need to take in order to enter frequencies into your radio will depend on the radio you've bought.

Today, many people purchase programming cables and software to get their radios programmed. They clone a friend's radio or get the frequencies programmed "at the store." But if you don't know how to enter frequencies directly from the front panel, you can run into problems in the field. You'll lose frequency agility—one of the advantages unique to ham radio operators, and one of the reasons that hams are valuable in emergency situations.

This guide has a simple purpose: to describe how to program handheld radios in the field. You will be able to program a frequency, CTCSS tone and offset into the radio, then transfer that programming to a memory so another radio operator can use it. If ham radio is deployed at an event or in an emergency, this may be enough to get you (or someone else) going when an existing repeater frequency has to change because of interference or range issues.

To be included in this guide, a radio must have programmable memories for frequency and CTCSS tones on transmit. Only radios which can be programmed in the field on their own—without using a computer or programming cable—are included.

Each entry has a standard format. First, there is a picture of the radio layout with interesting buttons marked. Then a few specifications for the radio are given (including transmit and receive frequency ranges). Next is a list of the standard tasks you might do with the radio. This includes programming a frequency to a memory in the field, locking/unlocking the radio, checking a repeater input frequency, changing power levels in the field, adjusting volume and adjusting squelch.

Many radios have modes that are easy to get into but hard to get out of. In the Weird Modes section, you may find assistance with this issue. At the end, you'll find other useful information about the radio, including how to reset it when nothing else works.

Entries are sorted lexicographically, not numerically or by date of manufacture. For instance, this means that the Yaesu FT-252 comes before the Yaesu FT-60.

1

The instructions provided for programming the radio may not be the most efficient. Instead, the instructions are as bulletproof as possible to users unfamiliar with the radio. For instance, with most of the Chinese radios, this guide has the user enter Menu mode, set the value and then exit Menu mode (which stores the setting). It's possible to chain multiple settings without leaving Menu mode on many of these radios. Feel free to take any shortcuts that your radio supports.

With many of the more modern radios, firmware updates are frequent. Menus might have moved around or changed number since the guide entry was written. If you don't see what you expect, try looking near that area for the appropriate function.

If your handheld isn't in the guide, look at other radios made by the same manufacturer at around the same time. Often manufacturers made multiple variants (e.g., 2m and 70cm) that were programmed identically.

Manufacturers usually sell similar radios for different IARU regions. Often there is an "A" version for the Americas (Region 2) and an "E" version for Europe (Region 1). Outside of the Americas, most of the 2m radios will transmit from 144–146 MHz; most of the 70cm radios will transmit from 430–440 MHz. Transmission on 1.25m is generally omitted on radios outside Region 2. Region 1 radios may have a button to transmit a 1750 Hz tone used to bring up European repeaters; sometimes this replaces the Monitor button. Radios in Region 2 have reception of mobile telephone frequencies blocked because of U.S. legislation.

Speaking of legalities, it's hard to talk about radios without using their names. All trademarks used in this book are properties of their respective owners.

⚠ **WARNING:** The information in this guide is intended to help you operate your radios in the field. It is provided "as-is" without any warranty. It's possible something in this guide may damage your radio. When in doubt, refer to the original user manual.

A Plea for Sanity

This book shouldn't exist. Radios today have enough horsepower and display capability to implement a standard method of programming. It seems as if there are two reasons this hasn't been done: the lack of a standard and lack of desire among radio manufacturers.

This book can't change the radio manufacturers, but it can suggest the following standard for radio programming.

All radios should have a button labeled Program. Pressing it should start a sequence of prompts to the user for all the parameters necessary for programming a memory:

1. the receive frequency

2. the transmit frequency

3. the transmit tone type (CTCSS, DCS or none)

4. the transmit tone value (skipped if there is no transmit tone type)

5. the receive tone type (CTCSS, DCS or none)

6. the receive tone value (skipped if there is no receive tone type)

7. the transmit power level

8. the memory location into which the values should be stored.

After the user enters the appropriate value, the Program button should move to the next step, until the memory is programmed. At that point, the radio should enter memory mode with the newly-added memory selected.

The user should be able to turn the radio off to abort programming. Radios with small displays should use scrolling text to prompt.

Today's radios are great pieces of hardware—but their firmware sometimes isn't quite as polished.

What is a VFO Anyway?

Amateur radio has a lot of terminology that may be unfamiliar to new people. Some terms come from the commercial radio industry, some from electronics, and some are unique to amateur radio.

It's not possible to explain it all—that would be a book in itself—but here are a few common words and phrases that are used in this book, with explanations for those who might not be in the know.

barrel connector Many handheld radios have a way to connect them to a power supply for charging. The most common connector used is a coaxial power connector, commonly called a barrel connector. These connectors have different inner and outer sizes.

C4FM A relatively new digital voice mode introduced in 2013 by Yaesu, proprietary and exclusive to some Yaesu radios. C4FM repeaters can be configured to receive both analog FM and digital C4FM mode, and transmit analog out.

center positive When a radio uses a coaxial power connector, it can be configured in one of two ways. The most common way is with positive voltage connected to the inner part of the connector, and negative voltage connected to the outer part of the connector. This is the center positive configuration. Some radios use a *center negative* configuration instead. Connecting a center positive connector to a center negative radio or vice versa can cause a fuse to blow or even damage the radio.

channel This is a term used by many Chinese radios for *memory*.

CTCSS tone Continuous Tone Coded Squelch System. Most repeaters will not retransmit audio unless they also receive a signal that indicates the transmission is intended for that repeater. This helps reduce unintended transmissions when two repeaters on the same frequency hear the same signal. A simple way to do this is to introduce a sub-audible tone which is played continuously while a radio transmits. Each repeater will then retransmit only when the appropriate tone is present. You might also hear this called a "PL" tone, which is a Motorola trademarked name for this system.

DCS Digital coded squelch. A more complex form of squelch than CTCSS. This system sends different tones in a digital pattern to open squelch, rather than a single tone. Occasionally called "CDCSS."

DMR Digital Mobile Radio. A commercial standard for digital voice which has been widely adopted commercially in North America, and adapted for amateur use. This mode has a standard, European Telecommunications Standards Institute TS 102 361.

D-STAR A digital standard for voice designed by the Japan Amateur Radio League and adopted by Icom. Recently, Kenwood has released a radio with D-STAR capabilities.

DTMF Dual-tone multiple frequencies. Many radios allow these touch-tone signals to be sent while pressing a number key during transmit. These tones can be used to control some repeaters.

dual watch Radios with a single receiver can receive only one frequency at a time. Some radios simulate two receivers by switching the single VFO rapidly between two frequencies. If a signal is detected, the radio stops switching frequencies and remains on the frequency with the signal.

Fusion See *C4FM*.

IARU region The International Amateur Radio Union defines three regions: Region 1 is Europe, Africa, Middle East and Northern Asia; Region 2 is the Americas; Region 3 is Asia Pacific. These regions conform to the International Telecom Union regions, so you might see them referred to as ITU regions.

lock Many radios offer the ability to prevent users from making unintentional changes to the configuration. Most offer a keyboard lock which causes the radio to ignore most keyboard buttons (with the notable exception of PTT). Some radios also offer a PTT lock and/or a knob lock, so the radio will ignore the PTT button or knob changes.

memory Modern radios offer a way to store and return to frequencies in a memory location on the radio. Often, radios will store other parameters in the memory, so they don't need to be set every time: shift, offset, squelch tones, sometimes power level. Memories may include a user-friendly name.

memory bank A way to group memories together. This is useful when scanning, and also to keep memories which are used for the same task within easy reach of each other.

offset When a radio is set to communicate through a repeater, it will be tuned to one frequency when transmitting, and then automatically retune itself to a different frequency when receiving. The difference in these two frequencies is the offset.

Part 90 The section of Federal Communications Commission regulations for commercial radios. In the United States, it is legal to use commercial radios in amateur bands.

power The strength with which a radio amplifies its transmit signal, usually measured in watts. Higher power levels will go longer distances, but use the battery faster. A radio's power level has no influence on how well it receives.

priority channel A special form of *dual watch*. Rather than switching between two arbitrary frequencies, the radio will switch between the current frequency and a previously-defined frequency. Radios will often check the priority frequency for activity even if there is activity on the other frequency.

PTT Push To Talk. A button on the radio which causes it to go into transmit mode when pressed, and into receive mode when released.

repeater A combination transmitter/receiver which retransmits what the receiver hears. Repeaters are often used for longer-distance communications between radios.

reset A way to revert the values in a radio to the original values before the radio was set up. Many radios offer ways to reset portions of the radio (a *partial reset*) rather than the entire radio (*full reset*).

step The minimum amount by which a frequency can change on the radio. For instance, a step size of 5 kHz might allow you to tune 147.000 MHz, 147.005 MHz, 147.010 MHz, 147.015 MHz.... Sometimes you will need to change the step size to select the frequency you want; you might have to use a step size of 6.25 kHz or 12.5 kHz to tune 147.0125 MHz.

shift In repeater operation, the shift determines if the offset is added to the receive frequency or subtracted from the receive frequency to yield the transmit frequency. A positive shift indicates the transmit frequency is greater than the receive frequency; a negative shift indicates the transmit frequency is less than the receive frequency.

simultaneous receive Some radios have two receivers built into the radio. These radios offer simultaneous receive: the ability to listen to transmissions on two frequencies at the same time.

squelch The ability of the radio to mute receive audio. *Carrier squelch* will mute the audio if no signal is present. *CTCSS squelch* will mute the audio if a subaudible tone is not present in the received signal. *DCS squelch* will mute the audio unless the received signal includes the correct digital data.

VFO Variable Frequency Oscillator. The component which alters the radio's frequency. Most handheld radios have one VFO; those with simultaneous receive have two. Fixed-frequency crystal-controlled radios do not have a VFO. Software-defined radios may simulate a VFO in software. Entering a frequency directly is often called "VFO mode."

Dead Battery Blues

Most early handhelds used nickel cadmium battery packs. These batteries discharged while on the shelf after about 30 days, and could handle about 500 charge cycles.

When the nickel metal hydride chemistry came along, users appreciated the greater energy densities. The improvement in energy density offset the disadvantage of a reduced number of charge cycles: NiMH cells can take about 300 charges before no longer being able to perform.

The transition from NiCd to NiMH was fairly smooth. Manufacturers found that NiMH batteries could be trickle-charged in the same way that NiCds could.

The advent of lithium ion batteries presented a problem to manufacturers. Li-ion batteries have a much higher power density than NiMH batteries, and have a lower self-discharge rate. They can be charged about the same number of times as NiMH batteries: 300 charge cycles. However, li-ion batteries need more advanced charging, and can't use the same trickle-charging circuitry that the nickel-based batteries did, which creates an incompatibility.

These days, almost all new radios use lithium ion batteries and have a charger designed for that. In the transition period from NiCd and NiMH to lithium ion, some manufacturers provided two chargers—one for the nickel chemistries, and one for the lithium ion chemistry. Then they made design changes to the battery packs to ensure the batteries were charged correctly. These changes were sometimes mechanical: additional plastic tabs and slots to prevent the battery from physically being inserted in the wrong charger, or charging pads at different locations than nickel-based batteries. Sometimes the changes were electronic: the radio had a menu to select between nickel batteries or lithium batteries and relied on the user to set it appropriately. In short, similar-looking battery packs and chargers may not work with each other.

There's not much you can do in the field if someone shows up with a dead battery. In rare cases, you might be able to rig a charger, *if* the radio has a charging jack or the user brought the correct charger.

Knowing about chargers for different battery chemistries may help a clueless user on future events. Possibly more helpful is a recommendation to buy an AA or AAA battery pack for the radio that can be used in place of the rechargable battery—and which will keep a radio powered long after the manufacturer stops production of the original batteries.

The Radio Has No Buttons

Before front panel programming, handheld radios had other methods of frequency entry. Some radios had no buttons at all. Even though radios like the Icom IC-2A had buttons on the front panel, they weren't for frequency entry—they just sent DTMF tones.

A radio before front panel programming. In this case the lack of an LCD screen and alternate function labels on the buttons suggest it doesn't have memories.

Setting the frequency

Early handheld radios had a single crystal-controlled frequency. There was no way to change those frequencies except by opening up the radio and swapping crystals. Slightly later, radios held more than one crystal, and frequencies were set via a switch (usually A/B or 1/2, sometimes A/B/C or 1/2/3 for radios that accommodated three crystals). Again, if the frequency you wanted wasn't one of the crystal frequencies, you'd have to open the radio up and swap crystals. If you're faced with a radio like this, the odds of getting it to work with a modern repeater are slim unless you routinely carry crystals around.

After the advent of phase lock loop (PLL) tuning, handheld radios had direct frequency entry, typically by setting thumbwheel switches on the top of the radio:

IC-2AT set to 144.390

These radios are almost always single band. You have to know which band the radio operates in. This is often but not always embedded in the model number. Because these radios are single band, users could assume the first two digits were "14" for 2m receivers, "22" for 220 MHz receivers, and "44" for 70cm receivers. Users then set the remaining digits using the thumbwheels.

On some radios, the thumbwheels simply add to the base frequency (144 on 2m, 220 on 220 MHz, and 440 on 70 cm). In these cases, you might set the thumbwheels to "039" to specify 144.390, and "324" to specify 147.240. The IC-2AT is more sophisticated: if the first thumbwheel is 0, 4 or 8 it uses 144 MHz; if the first thumbwheel is 1, 5 or 9 it uses 145 MHz; if the first thumbwheel is 2 or 6 it uses 146 MHz; if the first thumbwheel is 3 or 7 it uses 147 MHz. European variants which don't include 146–147 MHz will allow selection of 144–145 MHz only.

In addition to the thumbwheels, this radio has a +5 kHz switch. When this switch is turned off, the radio's frequency is what is set on the thumbwheel switches. When this switch is turned on, the radio tunes to 5 kHz higher than what the thumbwheels read (allowing a user to tune to 144.395 MHz, for instance).

Repeater operation

Radios which were made after repeaters became prevalent have another set of switches:

IC-2AT duplex and power settings

These switches determine whether the transmit frequency is different from the receive frequency, and if so by how much. (The output power level switch is also here for this transceiver.)

On radios that know about repeaters, the frequency you're setting using thumbwheel knobs is usually the receive frequency—but sometimes it's the transmit frequency.

For the most part, these radios do not allow CTCSS encoding. Some optional boards exist to provide tone encoding. Older boards transmit one CTCSS tone only. Newer boards can transmit different tones; they typically use DIP switches, jumpers, solder pads or a trim potentiometer to set the tone. Very recent tone boards include a digital display and buttons to set the tone.

Some tone boards have a separate potentiometer for CTCSS volume control.

A typical tone board with volume control   Solder pads set the frequency

When installed, tone boards are typically inside the body of the radio. You'll often need to open the radio to set the CTCSS tones, if you have that option.

I Don't Speak Chinese! (or Japanese, or Korean, or...)

Many radios have the ability to show their user interface in multiple languages. This allows manufacturers to use the same radio for different markets without having to flash different software for each market.

That's all fine until you actually have to reset the radio. At that point, the radio may revert to a language you don't speak. The smarter manufacturers have made the procedure for changing languages multilingual—for instance, some radios have both English and Chinese words for Language in their menu systems. Not everyone does that, though.

Even once you've found the "Language" section, you may hit a roadblock. Good systems show the word for "English" in English, the word for "Japanese" in Japanese, and the word for "Chinese" in Chinese. But some manufacturers take a different approach: if you're in English, the word for "Chinese" is just that: Chinese. That's no help to a native Chinese speaker. Similarly, if you're in Chinese, the word for "English" might be 英文寫作 (the Chinese characters that mean "English writing") rather than the word English.

Here are some common terms in different languages that you may encounter. If you have to reset a radio and it reverts to a non-English language, this may help you get back to something you can read. In general, the procedure is to look in the menu system for something that means "Language" and then (often one level below that in the menu system) something that means "English." Sometimes there is a menu above that—usually "System" or "Configuration."

If you're confronted with multiple possible options, it's best to write down which ones you have tried. This is easier to do if the menu system has numbers. If not, remember how many button pushes it took to get to that menu entry.

Chinese

In a Chinese menu system, you might look for a character included in "writing" 文 or a character included in "language" 语. From there you'd look for something including the character 英 (part of "English").

Japanese

In Japanese, you should look for something that has 語 in it (part of "language") and then look for something that has 英 (part of "English") in it. This character used in "English" is similar in Japanese and Chinese.

Korean

In Korean, you want something that includes at least one of the two characters 언어 in it ("language") and then look for something that looks like 영어 ("English").

Russian

In Russian, you'll see something like ЯЗЫК ("language"). Then look for английский ("English").

Other European Languages

If for some reason your radio defaults to Spanish, German, or French, look for a menu that contains "idioma," "Sprache" or "langue" ("language"). Then look for "Inglés," "Englische" or "Anglais(e)" ("English").

Radio Layout

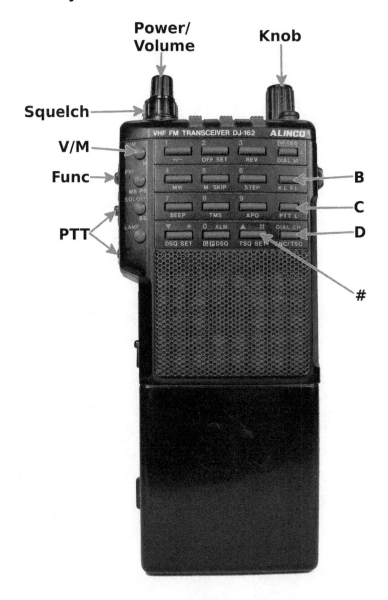

Specs

Receivers Single receiver
Receives 137.000–174.000 MHz FM
Transmits 144.000–147.995 MHz @ 5 W FM
Antenna connector BNC F on radio; needs BNC M antenna
Modes FM
Memory Channels 20
Power 9–13.8V DC, EIAJ-02 barrel style, 4mm OD, 1.7mm ID plug, **center negative**
Model year 1992

Standard Tasks

Program frequency in the field

1. Decide which memory to use. You can't overwrite—you will have to delete first if that memory has data in it. Start by pressing $\boxed{V/M}$ to enter memory mode if you aren't already there. (Screen displays M if you're in memory mode.) Use the knob to scroll to the memory you want to write to. If it has nothing in it, you will see a blinking M. If the M is solid, hold $\boxed{Func}$ and press $\boxed{4}$ to delete. Release $\boxed{Func}$.

2. Press $\boxed{V/M}$ to enter VFO mode. (Screen displays V.)

3. If you need to, adjust the step size. Hold $\boxed{Func}$ then press and release $\boxed{6}$ then rotate the knob to select from 0 50 (5 kHz), 1 (10 kHz), 1 25 (12.5 kHz), 2 (20 kHz) and 2 50 (25 kHz). Release $\boxed{Func}$ to exit.

4. Enter frequency (144390 for 144.390 MHz).

5. The radio has a global offset that applies to memories 0–16. To set it, hold $\boxed{Func}$ and press $\boxed{2}$. Release both buttons and use the knob to select the offset frequency in MHz. Press $\boxed{V/M}$ to set.

6. To set the shift, hold $\boxed{Func}$ and press $\boxed{1}$ to cycle from -, + and blank (no offset).

7. To set tone, hold $\boxed{Func}$ and press $\boxed{\#}$. Release $\boxed{\#}$ but keep $\boxed{Func}$ held down, and use the knob to select the tone. Release $\boxed{Func}$ to set.

8. To set the output power, slide the switch on the back of the radio. Sliding to the left (H) is high power (3 W); sliding to the right (L) is low power (0.3 W).

9. Press $\boxed{V/M}$ to enter memory mode. You will be on the memory to write.

10. Hold $\boxed{Func}$ then press $\boxed{4}$ to write to memory.

Lock/unlock radio

The radio has two locks: a PTT lock and a key/frequency lock. To lock or unlock PTT, hold Func and press C (display shows PTT when locked). To lock or unlock frequency or keyboard, hold Func then press B to cycle through KL (key lock), FL (frequency lock) or blank (unlocked).

Check repeater input frequency

Hold Func and press 3 to switch to reverse. Hold Func and press 3 to switch back.

Change power in the field

To set the output power, slide the switch on the back of the radio. Sliding to the left (H) is high power (3 W); sliding to the right (L) is low power (0.3 W).

Adjust volume

Use the inner part of the left knob (in front of the antenna jack) to adjust volume.

Adjust squelch

Use the outer ring of the left knob (in front of the antenna jack) to adjust squelch.

Weird Modes

Radio shows DSQ and/or transmits DTMF tones

This is not DCS as we know it. The radio has a pager mode that waits for specific DTMF tones, and sends DTMF tones on transmit. To disable, hold Func and press 0 repeatedly until DSQ disappears.

Radio doesn't transmit

This radio has a PTT lock. See the lock section to unlock.

Useful Information

This radio can have a tone squelch option installed (unit EJ-6U). If the radio has that option, you can enable tone squelch by holding Func and pressing D. (The display will show ENC when enabled.)

The radio has a DC power input; some batteries also have a DC power input. The battery inputs may have different ratings and/or polarities than the radio.

Memory channels 0 through 16 use the global value for offset if a shift is set. Memory channels 17 through 19 cannot have an offset.

Factory reset

Turn the radio on. Hold [Func] while turning the radio off and then back on again. This clears all memories and settings.

Radio Layout

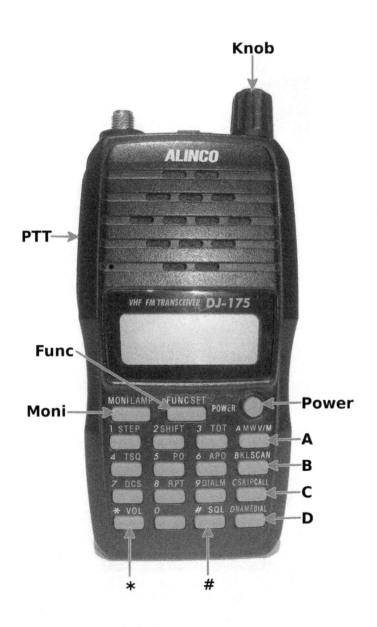

Specs

Receivers Single receiver
Receives 136.000–173.995 MHz FM
Transmits 144.000–147.995 MHz @ 5 W FM
Antenna connector SMA F on radio; needs SMA M antenna
Modes FM
Memory Channels 200
Power 12.0V DC, No DC input on radio
Model year 2008

Standard Tasks

Program frequency in the field

1. Decide which memory to use. You can't overwrite—you will have to delete first if that memory has data in it. Start by pressing [A] to enter memory mode if you aren't already there. (Screen displays M if you're in memory mode.) Press [Func] then scroll to the memory you want to delete. Press [A] to delete.

2. Press [A] to enter VFO mode if you aren't already there. (There will be no M on the screen if you're in VFO mode.)

3. Make sure your step size is set to 5. [Func] [1] then rotate the knob to select STP-5. Press [1] to exit.

4. Enter frequency (144390 for 144.390 MHz). (If step size is greater than 5, you might not have to / be able to enter the last digits.)

5. To set the shift, press [Func] [2]. Press [2] repeatedly to cycle from -0.600, +0.600 and OST-OF (no offset). You can adjust the shift if you need to by rotating the knob. Press [A] to set the value.

6. To set tone, press [Func] then [4]. Pressing [4] repeatedly cycles through T tone encoding, TSQ tone encoding and decoding, and nothing (no tone). Rotate the knob to select the tone. Press [A] to save. The radio can also do DCS using [Func] then [7].

7. To set the output power, press [Func] [5] to cycle through LO (0.5 W), MI (2 W) and high power (5 W—nothing displayed).

8. Press [A] to enter memory mode.

9. Press [Func]. Scroll to the desired memory (0–199). Memories with data will have a solid M; it will flash for empty memories. Make sure you're writing in an empty memory; if you pick an existing memory you will delete it instead.

10. Press [A] to write.

11. Press [A] to re-enter memory mode.

12. Scroll to the channel you just wrote using the knob.

Lock/unlock radio

Press Func B to lock the radio. Note that the keyboard can still send DTMF tones when locked. Press Func B to unlock.

Check repeater input frequency

There is no way to listen to the repeater input frequency except to program a separate memory for it.

Change power in the field

To set the output power, press Func 5. Rotate the knob to select from LO (0.5 W), MI (2 W) and high power (5 W—nothing displayed). Press 5 to set.

Adjust volume

Press *. Rotate the knob to set volume from 0–20. Press * to set.

Adjust squelch

Press #. Rotate the knob to set volume from 0–10. Press # to set.

Weird Modes

After a reset, volume and squelch are both 0. Change them to more reasonable values.

If the radio displays DISCHG, it is in battery refresh mode, which does a full discharge of the battery to reduce memory effect. Turn the radio off then on to leave this mode. (You can enter this mode by locking the radio, then pressing A A B B C C D D.)

Useful Information

The radio allows separate tones for encode / decode. Set one in T mode (encode) and a different one in TSQ (decode).

The radio has a "Set Mode" that allows you to set radio parameters. Enter it by holding Func for two seconds. Then you can scroll (using Func and Moni) through battery save BS, scan resume TIMER or BUSY, keyboard beep BEP, tone burst 1750 or another tone burst frequency, clock frequency shift SFT, busy channel lockout BCL, time after time out before you can transmit again TP, DTMF wait time DWT, DTMF pause DP, DTMF first digit time DB, and battery type BAT. Use the knob to adjust the value, and press A to save.

In most cases (except "Set Mode") if you wait five seconds after modifying a value, the value will be set to the new value automatically.

Factory reset

Turn the radio on while holding down [Func] and [A]. You don't get a chance to confirm this; the radio resets as soon as you release the keys.

Settings reset

Turn the radio on while holding [Func]. This resets everything except programmed memories. You don't get a chance to confirm this; the radio resets as soon as you release the keys.

Alinco DJ-180

Radio Layout

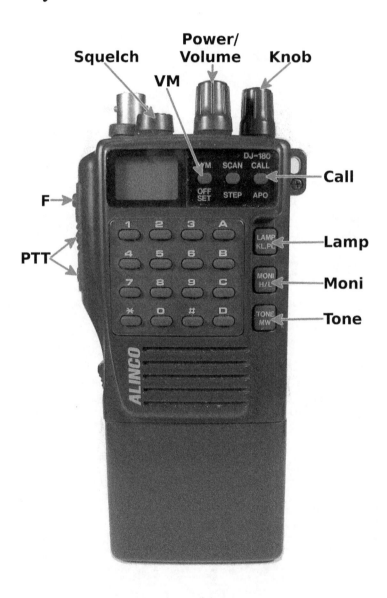

Squelch

VM

Power/Volume

Knob

Call

F

PTT

Lamp

Moni

Tone

Specs

Receivers Single receiver
Receives 130.000–173.995 MHz FM
Transmits 144.000–147.995 MHz @ 5 W FM
Antenna connector BNC F on radio; needs BNC M antenna
Modes FM
Memory Channels 9 standard, memory expansion options available for
    49 (EJ-14U) and 199 (EJ-15U)
Power 5.5–13.8V DC, No DC input on radio
Model year 1996

Standard Tasks

Program frequency in the field

1. Decide which memory to use. Make sure you're in memory mode (display shows M). Press V/M to get into memory mode if you're not there. Use knob to select memory to write to.

2. Press V/M to enter VFO mode. Use knob to select desired frequency. (Hold down F while using knob to change MHz.)

3. To set the tone, press Tone until display shows T (for tone on transmit) or TSQ (for tone on transmit and receive). Use knob to adjust tone value. Press V/M when done. You must have the EJ-17U CTCSS tone option installed for this to work.

4. To set the offset and shift, hold down F and press V/M. Use the knob to change the offset if required. Hold down F and press V/M to switch between no offset, - and +. Press V/M when you're done.

5. To set the power level, hold down F and press Moni. This will switch between H (high power = 5 W on 12 V battery, 2 W on 7.2 V battery) and L (low power = 0.4 W).

6. To write to the memory channel you have selected, hold down F and press Tone. This will write to the current memory. (You can press V/M and scroll to a different memory before writing if you need to; just press V/M again so you are in VFO mode when you write.)

7. Press V/M to enter memory mode.

8. Scroll to the channel you just wrote using the knob.

Lock/unlock radio

Hold F and press Lamp. You will cycle through KL (keyboard lock), PL (PTT lock), KLPL (both locked) and unlocked.

Check repeater input frequency

There is no way to listen to the repeater input frequency except to program a separate memory for it.

Change power in the field

Hold down [F] and press [Moni] to switch between high power (2–5 W depending on battery) and low power (0.4 W).

Adjust volume

Rotate center power/volume knob to adjust volume.

Adjust squelch

Rotate squelch knob on top left of radio.

Weird Modes

Radio displays AP

If the radio displays AP, it is in automatic power off mode. The radio will turn off if it goes 30 minutes with nothing breaking squelch. Turn the power knob off and back on again to turn the radio on again. Hold down [F] and press [Call] to enter/exit this mode.

Radio displays B

If the radio shows a B, the batteries are low. Change the batteries.

Useful Information

There are two [PTT] buttons on the left: the one marked PTT and the circular button below it. Both will transmit.

To enter extended frequency receive mode, turn radio off then turn it on while holding [Lamp]. Some versions of the radio require that you hold [F] and [Lamp] while turning it on.

Factory reset

Hold down [F] while turning on radio. This clears all memories.

Alinco DJ-190

Radio Layout

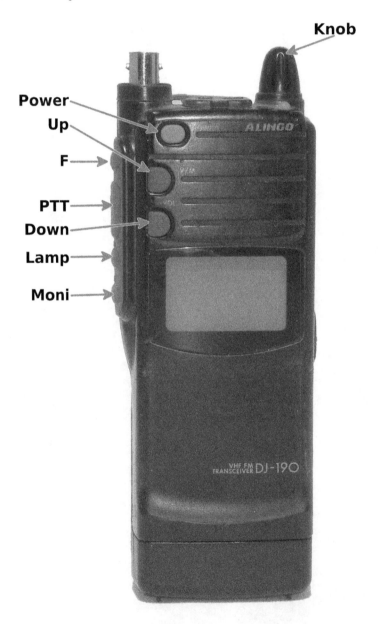

Specs

Receivers Single receiver
Receives 136.000–173.995 MHz FM
Transmits 144.000–147.995 MHz @ 5 W FM
Antenna connector BNC F on radio; needs BNC M antenna
Modes FM
Memory Channels 40
Power 4.8–13.8V DC, No DC input on radio
Model year 1997

Standard Tasks

Program frequency in the field

1. Decide which memory to use. Make sure you're in memory mode (display shows M). Hold $\boxed{F}$ and press $\boxed{Up}$ to get into memory mode if you're not there. Use knob to select memory to write to. (Flashing M is empty memory.)

2. Hold $\boxed{F}$ and press $\boxed{Up}$ to enter VFO mode. Use knob to select desired frequency. (Hold down $\boxed{F}$ while using knob to change MHz.)

3. To set the tone frequency, enter menu mode by holding down $\boxed{F}$ and pressing $\boxed{Moni}$. Press $\boxed{Up}$ or $\boxed{Down}$ until you see item 6, t NNN (where NNN is the current tone value). Use knob to adjust. Hold $\boxed{F}$ and press $\boxed{Moni}$ to exit.

4. To enable tone, hold $\boxed{F}$ and press $\boxed{Moni}$. Press $\boxed{Down}$ to menu item 5, t-SQL. Use knob to adjust. Display will show T if tone is enabled, or T SQL if tone squelch is enabled (requires EJ-28U CTCSS decoder to be installed). Hold $\boxed{F}$ and press $\boxed{Moni}$ to exit.

5. To set the offset, hold down $\boxed{F}$ and press $\boxed{Moni}$. Press $\boxed{Down}$ to go to menu 4, F N.NN (where N.NN is the offset frequency). Use knob to adjust. Hold down $\boxed{F}$ and press $\boxed{Moni}$ to exit.

6. To set the shift, hold down $\boxed{F}$ and press $\boxed{Moni}$. Press $\boxed{Down}$ to go to menu 3, Shift. Use knob to adjust. Hold down $\boxed{F}$ and press $\boxed{Moni}$ to exit.

7. To set the power level, hold down $\boxed{F}$ and press $\boxed{Lamp}$. This will switch between high power (blank): 5 W on 12 V battery, 1.5 W on standard 4.8 V battery) and L (low power = 0.8 W).

8. To write to the memory channel you have selected, hold down $\boxed{F}$ and press $\boxed{Down}$. This will write to the current memory. (You can hold $\boxed{F}$ and press $\boxed{Up}$ and scroll to a different memory before writing if you need to; just hold $\boxed{F}$ and press $\boxed{Up}$ again so you are in VFO mode when you write.)

9. Hold F and press Up to enter memory mode.

10. Scroll to the channel you just wrote using the knob.

Lock/unlock radio

Hold F and press Moni. Press Up or Down to go to menu 1, LoC. Use the knob to scroll between blank (unlocked), KL (key lock) and FL (frequency lock). Hold F and press Moni to set. To disable use the same procedure.

Check repeater input frequency

There is no way to listen to the repeater input frequency except to program a separate memory for it.

Change power in the field

Hold down F and press Lamp to switch between high power (1.5 W–5 W depending on battery) and low power (0.8 W).

Adjust volume

Press Up or Down to adjust volume from 0–31.

Adjust squelch

Hold down F and press Moni to enter menu mode. Press Up or Down to go to menu 16, SqLch. Rotate knob to adjust from 0–31. Hold F and press Moni to exit.

Weird Modes

Radio displays AP

If the radio displays AP, it is in automatic power off mode. The radio will turn off if it goes 30 minutes with nothing breaking squelch. Turn the radio on using the power button. To enter/exit this mode, hold down F and press Moni to enter menu mode. Press Up or Down to go to menu 7, APo. Use knob to adjust. (When enabled, display will show APO.) Hold down F and press Moni to exit.

Radio displays B

If the radio shows a B, the batteries are low. Change the batteries. (The switch on the back of radio releases the battery compartment.)

Radio displays ch- 0, **can't enter VFO mode**

This radio has a channel-only mode. When in this mode, you can only select pre-programmed memories. Display mode. To exit (or enter) it, first lock the radio (KL or FL). Then press and release ⬚ F ⬚. Then press ⬚Up⬚ six times, then ⬚Down⬚ three times. Finally, press ⬚Moni⬚.

Useful Information

Factory reset

Hold down ⬚ F ⬚ while turning on radio. This clears all memories.

Alinco DJ-480

Radio Layout

(Image is of DJ-180 which is the 2m variant)

Specs

Receivers Single receiver
Receives 430.000–459.995 MHz FM
Transmits 440.000–449.995 MHz @ 5 W FM
Antenna connector BNC F on radio; needs BNC M antenna
Modes FM
Memory Channels 9 standard, memory expansion options available for
    49 (EJ-14U) and 199 (EJ-15U)
Power 5.5–13.8V DC, No DC input on radio
Model year 1996

Standard Tasks

Program frequency in the field

1. Decide which memory to use. Make sure you're in memory mode
   (display shows M). Press [V/M] to get into memory mode if you're not
   there. Use knob to select memory to write to.

2. Press [V/M] to enter VFO mode. Use knob to select desired frequency.
   (Hold down [F] while using knob to change MHz.)

3. To set the tone, press [Tone] until display shows T (for tone on transmit) or TSQ (for tone on transmit and receive). Use knob to adjust tone
   value. Press [V/M] when done. You must have the EJ-17U CTCSS
   tone option installed for this to work.

4. To set the offset and shift, hold down [F] and press [V/M]. Use the
   knob to change the offset if required. Hold down [F] and press [V/M]
   to switch between no offset, - and +. Press [V/M] when you're done.

5. To set the power level, hold down [F] and press [Moni]. This will
   switch between H (high power = 5 W on 12 V battery, 2 W on 7.2 V
   battery) and L (low power = 0.4 W).

6. To write to the memory channel you have selected, hold down [F] and
   press [Tone]. This will write to the current memory. (You can press
   [V/M] and scroll to a different memory before writing if you need to;
   just press [V/M] again so you are in VFO mode when you write.)

7. Press [V/M] to enter memory mode.

8. Scroll to the channel you just wrote using the knob.

Lock/unlock radio

Hold [F] and press [Lamp]. You will cycle through KL (keyboard lock), PL
(PTT lock), KLPL (both locked) and unlocked.

Check repeater input frequency

There is no way to listen to the repeater input frequency except to program a separate memory for it.

Change power in the field

Hold down F and press Moni to switch between high power (2–5 W depending on battery) and low power (0.4 W).

Adjust volume

Rotate center power/volume knob to adjust volume.

Adjust squelch

Rotate squelch knob on top left of radio.

Weird Modes

Radio shows AP

If the radio displays AP, it is in automatic power off mode. The radio will turn off if it goes 30 minutes with nothing breaking squelch. Turn the power knob off and back on again to turn the radio on again. Hold down F and press Call to enter/exit this mode.

Radio shows B

If the radio shows a B, the batteries are low.

Useful Information

There are two PTT buttons on the left: the one marked PTT and the circular button below it. Both will transmit.

To enter extended frequency receive mode, turn radio off then turn it on while holding Lamp . Some versions of the radio require that you hold F and Lamp while turning it on.

Factory reset

Hold down F while turning on radio. This clears all memories.

Alinco DJ-500T Gen 2

Radio Layout

Specs

Receivers Single receiver, dual watch (first to break squelch wins)
Receives 136–174 MHz FM, 400–480 MHz FM
Transmits 144–148 MHz @ 5 W FM, 420–450 MHz @ 5 W FM
Antenna connector SMA F on radio; needs SMA M antenna
Modes FM
Memory Channels 200
Power 7.4V DC, No DC input on radio
Model year 2014

Standard Tasks

Program frequency in the field

1. Press C/Down to enter VFO mode if you're not already there (look for a number in the lower right; that indicates you're in memory mode and need to switch).

2. Enter the frequency on the keypad (144390 for 144.390 MHz).

3. To set the transmit tone, press A 8 to enter set menu. Use B/Up and C/Down to scroll to menu 01, T-CDC. Press 1 until the display shows CT (other choices are DCS and OFF). Use knob to adjust tone value. Press D when done. You can also adjust menu 02, R-CDC for receive tones if desired.

4. To set the offset, press A 8 to enter set mode. Scroll to menu 11, OFFSET using B/Up / C/Down. Use the knob to adjust the offset. Press D when done.

5. To set the shift, press A 4. Repeatedly press 4 to cycle through -, + and blank (no offset). Press A when done.

6. To set the power level, press A 9. Repeatedly press 9 to cycle through H (high power = 5 W), M (medium power = 2.5 W) and L (low power = 1 W). Press A when done.

7. To write to the memory channel you have selected, press A then C/Down and the channel number will blink. Use the knob to select the desired channel. Hold C/Down until you hear two beeps.

8. Press C/Down to enter memory mode.

9. Scroll to the channel you just wrote using the knob.

Lock/unlock radio

Press A then hold # for two seconds to lock or unlock. After unlocking, ensure you are out of function mode (F is not visible) by pressing A if necessary.

Check repeater input frequency

Press [A] [7] [A] to enter reverse mode (display shows R). Repeat to exit.

Change power in the field

Press [A] [9]. Repeatedly press [9] to cycle through H (high power = 5 W), M (medium power = 2.5 W) and L (low power = 1 W). Press [A] when done.

Adjust volume

Rotate right power/volume knob to adjust volume.

Adjust squelch

Press [A] [8] to enter the set menu. Scroll to menu 23 SQL using [B/Up] and [C/Down]. Use the knob to adjust squelch from 00 (open) to 09. Press [D] to exit.

Weird Modes

Can't enter VFO mode

⚠ **WARNING:** It is possible to enter channel mode from the keypad. Once you're in channel mode, there is *no way* to get back to VFO mode without a computer and cable. This mode is entered by holding down [PF2] while turning on the radio, then pressing [B/Up] / [C/Down] to go to menu 01, DSP. If you enable CH and press [#] you will be unable to program the VFO. This mode survives the factory reset.

Radio does not transmit

This radio has a transmit inhibit function. To disable, press [A] [8] to enter the set menu. Scroll to menu 14 TX using [B/Up] and [C/Down]. Use the knob to switch to ON to enable transmit, or OFF to disable transmit. Press [D] to exit.

Radio button selects wrong mode

The function mode F enabled by pressing [A] is a toggle. The radio will remain in F mode until you press [A] again.

Radio doesn't switch frequencies for repeater

This radio has a talk-around mode which keeps the radio on the output frequency. To enable/disable it, press [A] [8] to enter the set menu. Scroll to menu 10 TALKAR using [B/Up] and [C/Down]. Use the knob to switch to OFF to disable talkaround, or ON to enable it. Press [D] to exit.

Useful Information

Alinco made two generations of the DJ-500. These instructions are appropriate for the second generation with bitmap display.

Factory reset

To reset everything, hold down PF2 while turning on radio to enter special set menu. Use B/Up and C/Down to select menu 02, RESTOR. Use knob to select FACT?. Press # to reset (immediately; no prompt).

Settings reset

To reset settings but leave memories intact, hold down PF2 while turning on radio to enter special set menu. Use B/Up and C/Down to select menu 02, RESTOR. Use knob to select INIT?. Press # to reset (immediately; no prompt).

Radio Layout

Specs

Receivers Single receiver
Receives 144.000–147.995 MHz FM, 420.000–449.995 MHz FM
Transmits 144.000–147.995 MHz @ 4.5 W FM, 420.000–449.995 MHz @ 4 W FM
Antenna connector BNC F on radio; needs BNC M antenna
Modes FM
Memory Channels 100
Power 6.0–16.0V DC, charger is 13.8V DC, "H" barrel style jack 3.4mm OD, 1.3 mm ID center positive. Note that DJ-596T Mk II is 6.0–15.0V DC
Model year 2001

Standard Tasks

Program frequency in the field

1. Decide which memory to use. You can't overwrite—you will have to delete first if that memory has data in it. Start by pressing ⏎A⏎ to enter memory mode if you aren't already there. (Screen displays M if you're in memory mode.) Press ⏎Func⏎ then scroll to the memory you want to delete. Press ⏎A⏎ to delete.

2. Press ⏎A⏎ to enter VFO mode if you aren't already there. (There will be no M on the screen if you're in VFO mode.)

3. Press ⏎Func⏎ ⏎D⏎ to change band if necessary.

4. Make sure your step size is set to 5. ⏎Func⏎ ⏎1⏎ then rotate the knob to select STP-5. Press ⏎1⏎ to exit.

5. Enter frequency (144390 for 144.390 MHz). (If step size is greater than 5, you might not have to / be able to enter the last digits.)

6. To set the shift, press ⏎Func⏎ ⏎2⏎. Press ⏎2⏎ repeatedly to cycle from -, +, SPLIT (odd splits) and OST-OF (no offset). You can adjust the shift if you need to by rotating the knob. (Press ⏎Func⏎ to increase the adjustment factor.) Press ⏎A⏎ to set the value.

7. To set tone, press ⏎Func⏎ then ⏎4⏎. Pressing ⏎4⏎ repeatedly cycles through T tone encoding, TSQ tone encoding and decoding, and TCS-OF (no tone). Rotate the knob to select the tone. Press ⏎A⏎ to save. The radio can also do DCS using ⏎Func⏎ then ⏎7⏎.

8. To set the output power, press ⏎Func⏎ ⏎5⏎ to cycle through LO (0.8 W) and high power (2.5–5 W depending on battery—nothing displayed).

9. Press ⏎A⏎ to enter memory mode.

10. Scroll to the desired memory (0–99). Memories with data will have a solid M; it will flash for empty memories. Make sure you're writing in an empty memory; if you pick an existing memory you will delete it instead.

11. Press ⬚Func⬚ ⬚A⬚ to write.

Lock/unlock radio

Press ⬚Func⬚ ⬚B⬚ to lock the radio. Note that the keyboard can still send DTMF tones when locked. Press ⬚Func⬚ ⬚B⬚ to unlock.

Check repeater input frequency

There is no way to listen to the repeater input frequency except to program a separate memory for it.

Change power in the field

To set the output power, press ⬚Func⬚ ⬚5⬚. Rotate the knob to select from LO (0.8 W) and high power (2.5–5 W depending on battery—nothing displayed). Press ⬚5⬚ to set.

Adjust volume

Press ⬚*⬚. Rotate the knob to set volume from 0–20. Press ⬚*⬚ to set.

Adjust squelch

Press ⬚#⬚. Rotate the knob to set squelch from 0–20. Press ⬚#⬚ to set.

Weird Modes

After a reset, volume and squelch are both 0. Change them to more reasonable values.

Useful Information

The radio allows separate tones for encode / decode. Set one in T mode (encode) and a different one in TSQ (decode).

The radio has a "Set Mode" that allows you to set radio parameters. Enter it by holding ⬚Func⬚ for three seconds. Then you can scroll (using ⬚Func⬚ and ⬚Moni⬚) through battery save BS, scan resume TIMER or BUSY, keyboard beep BEP, tone burst 1750 or another tone burst frequency, busy channel lockout BCL, time after time out before you can transmit again TP, DTMF wait time DWT, DTMF pause DP, DTMF first digit time DB, theft alarm SCR (using a speaker plug where ring and ground are connected), +5 V DC on mic plug during transmit EXP, ultrasonic mosquito repellent(!) MRS and roger beep EDP. Use the knob to adjust the value, and press ⬚A⬚ to save.

Factory reset

Turn the radio on while holding down ⌈Func⌉. You don't get a chance to confirm this; the radio resets as soon as you release the keys.

Alinco DJ-C5

Radio Layout

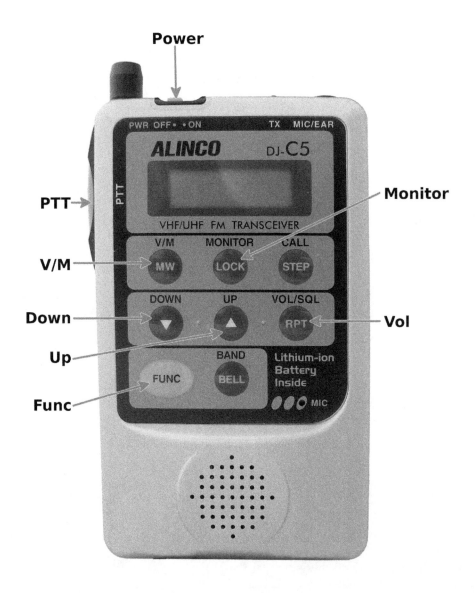

Specs

Receivers Single receiver
Receives 118.000–135.995 MHz AM, 136.000–173.995 MHz FM, 420.000–449.995 MHz FM
Transmits 144.000–147.995 MHz @ 300mW FM, 420–449.995 MHz @ 300 mW FM
Antenna connector Custom connector on radio and antenna
Modes FM
Memory Channels 50
Power 4.2V DC (measured) to custom two-pin charge port on radio; 9V DC, IEC 60130-10 type C 3.8mm OD/1.4mm ID center positive to charger
Model year 1998

Standard Tasks

Program frequency in the field

1. Decide which memory to use. Make sure you're in memory mode (display shows M). Press [V/M] to get into memory mode if you're not there. Press [Func] then [Up] and [Down] to scroll to the memory you want.

2. Press [V/M] to enter VFO mode. Press [Band] to choose band. Use [Up] / [Down] to select desired frequency. (If you press [Func] before pressing [Up] and [Down], MHz frequency will change.)

3. To set the tone, hold [Func] for one second, then [Vol] until the display shows a non-blinking T. Press [Up] / [Down] to select tone. Press [PTT] (does not transmit) to exit.

4. To set the offset and shift, press [Func] and then [Vol]. Press [Vol] repeatedly to change from -, + no offset. Use [Up] and [Down] to change the offset if required. Press [PTT] (does not transmit) when you're done.

5. Power level is fixed at 300 mW.

6. To write to the memory channel you have selected, press [Func] then press [V/M]. This will write to the current memory.

7. Press [V/M] to enter memory mode.

8. Scroll to the channel you just wrote using [Up] and [Down].

Lock/unlock radio

Press [Func] then [Monitor]. This will lock the keyboard, but will still allow PTT and volume/squelch adjustments.

Check repeater input frequency

Press Monitor while offset is on. Press Monitor again to get back to normal.

Change power in the field

This radio has a fixed 300 mW power level.

Adjust volume

Press Vol. Display will show VoL. Press Up / Down to adjust volume (0–8). Press PTT (does not transmit) to set.

Adjust squelch

Press Vol twice. Display will show SqL. Press Up / Down to adjust squelch (0–5). Press PTT (does not transmit) to set.

Weird Modes

Frequency not displayed

The radio can be put in a channel mode. If that happens, it will display CH . 1. To disable or re-enable channel mode, hold down V/M while turning the radio on.

Useful Information

The radio uses a custom antenna connector that appears to be a 2.95mm diameter threaded rod.

Custom antenna connector

Tone squelch on this radio does not provide all frequencies. You can set tone squelch from 67.0–156.7 only.

If you hold down Up or Down while not adjusting a value, the radio will start to scan. Press Func to stop it.

Radios which have been modified for extended range will cycle through three values for band: 145.00, 380.00 and 445.00 (by default).

You can adjust the step size. Press Func then Call. Press Up / Down to adjust, then PTT (does not transmit) to set.

Factory reset

Hold down Func and V/M while turning on radio, then release buttons. This clears all memories.

Radio Layout

Specs

Receivers Two independent receivers, simultaneous receive. Transmit on Main only, not Sub.

Receives 216–249.995 MHz and 902–927.995 MHz FM

Transmits 222–224.995 MHz @ 4 W FM and 902–927.995 MHz @ 1.7 W FM

Antenna connector SMA F on radio; needs SMA M antenna

Modes FM

Memory Channels 100 in each of 5 banks (Bank 0–Bank 4)

Power 9–16V DC, IEC 60130-10 type E 3.4mm OD/1.3mm ID center positive

Model year 2011

Standard Tasks

Program frequency in the field

1. Decide which memory to use. You can't overwrite—you will have to delete first if that memory has data in it. You must enable deletion if it isn't already enabled on the radio.

2. To enable deletion, press ⎣Func⎦⎣Lamp⎦⎣Lamp⎦ then rotate Right Knob to <MEMORY>. Push ⎣Right Knob⎦. Rotate Right Knob to Overwrite. Rotate Right Ring to fail-safe (allows deletion until radio is turned off) or Accepted (allows deletion always). Press ⎣Func⎦.

3. To delete a frequency, Press ⎣V/P/M⎦ until you're in memory mode. In memory mode, press ⎣Main⎦ repeatedly to go through the banks until you reach the bank you want to delete from. Use the Left Knob to select the memory in the bank that you want to delete. Press ⎣Func⎦ ⎣Clr⎦⎣Ent⎦ to delete that memory.

4. Press ⎣V/P/M⎦ until you're in VFO mode. Enter frequency (9270500 for 927.0500). You may need to press ⎣Func⎦ ⎣7⎦ to adjust step size to enter the frequency. (Rotate Left Knob to adjust step, then press ⎣Func⎦ again when done.)

5. To set the tone, press ⎣Func⎦ ⎣5⎦ to enter tone set mode. Press ⎣5⎦ to cycle through Tx TSQ Rx OFF, Tone Squelch, Tx TSQ Rx TSQrev, Tx TSQ Rx DCS, Tx DCS Rx OFF, Tx DCS Rx TSQ, Tx DCS RX TSQrev, DCS, Tx OFF Rx TSQ, Tx OFF Rx TSQrev, Tx OFF Rx DCS and OFF. Rotate Left Knob to set the encoding value. Rotate Right Knob to set the decoding value (if applicable). Note that earlier versions of the firmware (before 1.10) don't have all of these options. Press ⎣Func⎦ to set.

6. This radio has automatic repeater shift hard coded for 223.91–225 MHz (-1.6 MHz) and 927–927.995 (-25 MHz). See "Useful Information" below for instructions to turn automatic repeater shift off and

enable setting the repeater shift manually. To set the repeater shift once ARS has been turned off, press [Func] [Main] to enter shift mode. Press [Main] to cycle through -, + and OFF (no offset). Rotate the Left Knob to select the value. Press [Func] when done.

7. To set the power level, press [Func] [2]. Use Left Knob to select from Low Power, Middle Power and High Power. Press [Func] when done.

8. To write to a memory channel, press [Func] then rotate Left Ring to select the band. (Bands 0 through 4 are "Regular" memories.) Rotate the Left Knob to select the memory slot in the band. Press [V/P/M] to write.

9. Press [V/P/M] to enter memory mode.

10. Scroll to the channel you just wrote. You may need to press [Main] to change bands.

Lock/unlock radio

Hold [Func] for two seconds to quick lock/unlock (black key symbol on white background).

Hold [Sub] while pressing [Left Knob] three times for normal lock/unlock (white key symbol on black background).

Check repeater input frequency

Press [Ent]. Split indicator will flash to show you're in reverse mode. Press [Ent] again to return to normal mode.

Change power in the field

Press [Func] [2] (PO) then use either knob to adjust from Low Power (0.3 W), Medium Power (0.8 W) and High Power (1.7 W on 902 MHz, 4 W on 222 MHz). Press [Func] to set.

Adjust volume

For main (transmit) band, rotate outer ring of Left Knob to adjust volume.

For sub (receive only) band, rotate outer ring of Right Knob to adjust volume.

Adjust squelch

For main (transmit) band, press [Left Knob], then rotate Left Knob to desired squelch. Press [Left Knob] to set.

For sub (receive only) band, press [Right Knob], then rotate Right Knob to desired squelch. Press [Right Knob] to set.

Weird Modes

The V/P/M button doesn't work and there's a horizontal line on the screen

You're probably in scope mode. Press [Func] [6] to leave it.

Unable to set repeater shift between 223.91–225 MHz or between 927–927.995 MHz

Turn the automatic repeater shift off as detailed in "Useful Info."

Useful Information

The Left Knob and Right Knob on this receiver also act as buttons when pushed.

To turn power on, hold [Power] for at least one second. Some users have reported the power button is stiff, so press it firmly.

This receiver has at least two firmware versions. Version T1.00 cannot do DCS on transmit and CTCSS on receive or vice versa. Version T1.10 can.

To determine the firmware version of your receiver, hold [Func] for two seconds to quick lock. Then press [1] ten times to display the firmware version. Press [Func] to clear.

When automatic repeater shift has been turned on, you are prevented from setting the repeater shift for certain ranges. Frequencies between 223.91 and 225 MHz are set to -1.6 MHz, and frequencies between 927 and 927.995 MHz are set to -25 MHz (1.10 firmware). To disable automatic repeater shift and allow you to set the repeater shift manually, press [Func] [Lamp] [Lamp], then rotate Right Knob to <REPEATER>. Press [Right Knob]. Rotate Right Ring until Auto rpt. Set is OFF. Press [Func] to set.

Factory reset

Hold down [7], [Scan] and [Func] while turning radio on. This resets everything including memories.

Settings reset

Hold down [Func] while turning radio on. This resets the radio to default settings, but preserves memories.

Radio Layout

Specs

Receivers Single receiver
Receives 144.000–147.995 MHz MHz FM
Transmits 144.000–147.995 MHz @ 340 mW FM
Antenna connector Swing-out five-segment telescoping whip (built-in);
    no provision for alternate antenna
Modes FM
Memory Channels 20
Power 5.5 V DC, EIAJ-02 connector, 4mm OD, 1.65mm ID, center positive
Model year 1996

Standard Tasks

Program frequency in the field

1. Make sure you are in VFO mode. If you're in memory mode (display
   shows M) then press [V/M] once.

2. Use [Up] and [Down] to select desired frequency. For faster changes,
   press and release F, to adjust the values for MHz. Press and release
   [F] again to adjust 100 kHz.

3. To set the tone, hold down [F] and press [Down] (Tone). Use [Up] and
   [Down] to select tone. Hold down [F] and press [Down] (Tone) while in
   this mode to turn tone on (display shows T) or off. Press [PTT] when
   done (does not transmit). This radio has tone encoding only, not tone
   squelch.

4. To set the offset and shift, hold down [F] and press [Scan] (Shift).
   Use up and down to adjust the offset (0.60 is 600 kHz). Press [Scan]
   while in this mode to switch between - offset, + offset and no offset.
   Press [PTT] when done (does not transmit).

5. To set the power level, hold down [PTT] and press [Scan]. This will
   switch between high power (340 mW) and low power (50 mW). Adjusting the power in this way **does transmit**.

6. To write to the memory channel you have selected, hold down [F] and
   press [V/M] (MW). Then use [Up] and [Down] to select the memory you
   want to write to (0–19).

7. Press [V/M] to write that memory.

8. Press [V/M] to enter memory mode.

9. Use [Up] and [Down] to go to the memory you just wrote.

Lock/unlock radio

Hold [F] and press [Moni]. The radio will display an L. Hold [F] and press
[Moni] again to unlock.

Check repeater input frequency

There is no way to listen to the repeater input frequency except to program a separate memory for it.

Change power in the field

Hold down $\boxed{\text{PTT}}$ and press $\boxed{\text{Scan}}$. This will switch between high power (340 mW) and low power (50 mW). Adjusting the power in this way **does transmit**.

Adjust volume

Rotate power/volume knob to adjust volume.

Adjust squelch

Squelch is fixed and cannot be adjusted.

Weird Modes

Radio displays OFF instead of transmitting

The radio is trying to transmit beyond the band edge. Check your offset.

Radio plays courtesy beep after transmission

Hold down $\boxed{\text{Down}}$ and turn the radio on to disable courtesy beep. Hold down $\boxed{\text{Up}}$ and turn the radio on to re-enable.

Can't enter VFO mode; display shows ch XX

You are in channel-only mode. To enter or exit this mode, hold down $\boxed{\text{V/M}}$ while turning the radio on.

Useful Information

Hold down $\boxed{\text{Moni}}$ and turn the radio on to enable/disable key beep.

The radio takes three AA batteries. Do not connect any external power to the radio while there are batteries in the radio.

Always extend the antenna when transmitting. Be careful; replacement antennas are no longer available.

Factory reset

Hold down $\boxed{\text{F}}$ while turning on radio. This clears all memories and settings.

Alinco DJ-V5

Radio Layout

Specs

Receivers Single receiver with priority channel option
Receives 144.000–147.995 MHz FM, 420.000–449.995 MHz FM
Transmits 144.000–147.995 MHz @ 6 W FM, 420.000–449.995 MHz @ 6 W FM
Antenna connector SMA F on radio; needs SMA M antenna
Modes FM
Memory Channels 200
Power 4.0–15.0V DC, charger is 13.8V DC, "H" barrel style jack 3.4mm OD, 1.3 mm ID center positive
Model year 2000

Standard Tasks

Program frequency in the field

1. Press ⟨A⟩ (V/M) to enter VFO mode if you aren't already there. (There will be no M on the screen if you're in VFO mode.)

2. Press ⟨Band⟩.

3. Make sure your step size is set correctly. Press ⟨Func⟩ then ⟨C⟩ (STEP), and then rotate the knob to select from 5.0K, 10.0K, 12.5K, 20.0K, 25.0K, 50.0K, 100.0K. Press ⟨PTT⟩ (does not transmit) to set.

4. Enter frequency (144390 for 144.390 MHz). (If step size is greater than 5.0, you might not have to / be able to enter the last digits.)

5. To set the shift, press ⟨D⟩ (RPT) repeatedly to cycle from -, +, and no display (simplex). You can adjust the shift if you need to by rotating the knob. Press ⟨PTT⟩ to set (does not transmit).

6. To set tone, press ⟨Func⟩ then ⟨8⟩ (T SQL). Pressing ⟨8⟩ repeatedly cycles through T tone encoding, T SQ tone encoding and decoding, and OFF (no tone). Rotate the knob to select the tone. Press ⟨PTT⟩ to set (does not transmit).

7. To set the output power, press ⟨Func⟩ ⟨6⟩ (PO). Press ⟨6⟩ repeatedly to cycle through HI (6.0 W), L1 (1.0 W) and L2 (0.5 W). Press ⟨PTT⟩ to set (does not transmit). Note that L1 is higher power than L2.

8. Press ⟨Func⟩ to enter memory writing mode.

9. Scroll to the desired memory (0–199) using the knob. (Avoid other values which represent band edges and call frequencies.) Memories with data will have a solid M; it will flash for empty memories.

10. Press ⟨A⟩ (MW) to write.

Lock/unlock radio

Hold ⟨Func⟩ for one second to lock the radio. Left side keys will still work when locked. Hold ⟨Func⟩ for one second to unlock.

Check repeater input frequency

Hold down [0] (REV) to monitor input frequency. Release the key to return to normal operation.

Change power in the field

To set the output power, press [Func] [6] (PO). Press [6] repeatedly to cycle through HI (6.0 W), L1 (1.0 W) and L2 (0.5 W). Press [PTT] to set (does not transmit). Note that L1 is higher power than L2.

Adjust volume

Rotate the ring to set volume.

Adjust squelch

Hold down [Moni] and rotate the knob to choose from L0 (open squelch) through L5. Release [Moni] to set.

Weird Modes

Can't enter frequencies

This radio has a "call mode" which prevents you from entering frequencies. When in call mode, the display will show C1 or C2. Press [A] (V/M) to go into memory mode from call mode.

Radio switches frequency every five seconds

The radio has a priority watch. If it is on, the LCD will show PRIO and will check the memory that was selected when priority was turned on. Press [Func] then [1] (PRIO) to turn it off/on.

Memories display as CH nn

To switch from channel display mode to frequency display mode, hold down [A] while turning on power. Repeat operation to switch back.

Useful Information

The radio allows separate tones for encode / decode. Set one in T mode (encode) and a different one in TSQ (decode).

Note that DSQ ([Func] then [7]) is DTMF squelch, not digital coded squelch.

The radio has a "Set Mode" that allows you to set radio parameters. Enter it by pressing [Func] then [Band]. Then you can scroll (using [*] (down) and [#] (up) keys) through BP- (key beep), BEL (beep when DTMF squelch tone is received), APO (auto power off), BS- (battery save), DT-

(DTMF squelch delay), and SP- (split). Adjust values using the knob, and press [PTT] to save.

This radio has a one-wire mod to enable multi-band receive. If that mod has been done, receive range is 76.0–999.995 MHz (cell excepted). This mod also enables extended transmit, so be careful not to program out of band. You can check by pressing [Band] repeatedly—if you cycle through bands including 380.0 MHz and 800 MHz, the mod has been performed.

Factory reset

Turn the radio on while holding down [Band]. You will be prompted with RESET*. Press [*] to reset.

Azden AZ-21A

Radio Layout

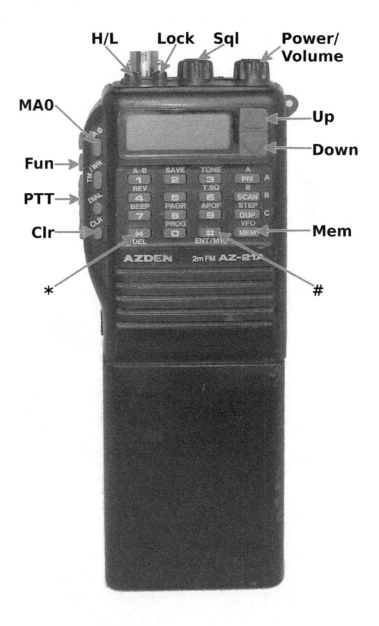

Specs

Receivers Single receiver
Receives 118.000–173.995 MHz FM
Transmits 144.000–147.995 MHz @ 5 W FM
Antenna connector BNC F on radio; needs BNC M antenna
Modes FM
Memory Channels 21
Power 12.0V DC, EIAJ-03 barrel style, 4.7mm OD, 1.7mm ID plug, center
     positive
Model year 1993

Standard Tasks

Program frequency in the field

1. This radio requires that you program separate transmit and receive
   frequencies—not a receive frequency and an offset. Determine ahead
   of time what the transmit and receive frequencies are and have them
   ready.

2. There is a ten-second timeout. If you exceed this while programming,
   you will have to start again. Write down what you are going to do
   before you do it.

3. The [*] is used for a decimal key.

4. Hold [Fun] and [0] for one second to enter programming mode. Display will show PR.

5. Use [Up] and [Down] to select memory (MA0 or 01–20). Press [#].

6. Enter receive frequency in MHz, using [*] for the decimal point.
   147*240 for 147.240 MHz. Press [#].

7. Use [Up] and [Down] to select receive CTCSS. C00: means disable receive squelch. Press [#].

8. Enter transmit frequency in MHz, using [*] for the decimal. 147*840
   for 147.840 MHz. Press [#].

9. Use [Up] and [Down] to select transmit CTCSS. C00: means disable
   receive squelch. Press [#].

10. Output power cannot be set for a specific memory.

11. Wait ten seconds for the memory to time out.

12. When you first program a memory, it is disabled. Press [Mem] and
    then very quickly enter the memory you just wrote followed by [#]
    (02# for memory 02).

13. Press [#] to enable the memory. The << next to the channel number
    will disappear. (You can disable again with [*].)

Lock/unlock radio

Press [Lock] to lock or unlock the radio. The button latches; up is unlocked.

Check repeater input frequency

Hold [Fun] and press [4] to go to reverse mode. Display shows REV. Repeat
the sequence to go back to normal operation.

Change power in the field

Press [H/L] to switch between low (0.5 W) and high (5.0 W) power. The
button latches; up is high power. Display will show LO when in low power
mode.

Adjust volume

Rotate the right knob to set volume.

Adjust squelch

Rotate the left knob to set squelch.

Weird Modes

None known.

Useful Information

The MA0 memory can be retrieved by pressing [MA0].
    To disable the keyboard beep, turn the radio on while holding [Fun].

Factory reset

To reset the radio, turn it on while holding [Clr].

Baofeng UV-100 / UV-200 / UV-3R

Radio Layout

A UV-200 model

Specs

Receivers Single receiver, dual watch (first to break squelch wins)
Receives 136–174 MHz and 400–470 MHz FM
Transmits 136–174 MHz @ 2 W FM and 400–470 MHz @ 2 W FM
Antenna connector SMA **M** on radio; needs SMA **F** antenna
Modes FM
Memory Channels 99
Power No DC input on radio; 5 V DC, 2.4 mm OD, 0.70 mm ID, **center negative** on charger base
Model year 2011

Standard Tasks

Program frequency in the field

1. Hold U/V for two seconds to go to VFO mode if you aren't there already.

2. Depending on the frequency you want to program, it may be necessary to adjust the step beforehand. To do this, Press Menu and scroll to 09 - STEP to set. Press U/V to adjust and then scroll with the knob to the correct step (5, 6.25, 10, 12.5, 20, 25). Press Menu to set.

3. Use the knob to select frequency. Enter UHF frequencies on the top and VHF frequencies on the bottom. Press U/V to switch between top and bottom. To make things quicker, you can press F/A before turning the knob to set MHz. (Press F/A again to leave the MHz set mode.)

4. Press Menu and scroll to 02 - TXCODE to set CTCSS / DCS. Press U/V to adjust and then scroll with the knob to the correct value (numbers with a decimal point are CTCSS; values ending in I or N are DCS). Press Menu to set.

5. Press Menu and scroll to 10 - OFFSET to set repeater offset. Press U/V to adjust and then scroll with the knob to the correct value. When adjusting you can press F/A to adjust MHz; press F/A again to go back to adjusting kHz. Press Menu to set.

6. Press Menu and scroll to 11 - SHIFT to set repeater shift. Press U/V to adjust and then scroll with the knob to the correct value (0, -, +). Press Menu to set.

7. Press Menu and scroll to 07 - POWER to set power level. Press U/V to adjust and then scroll with the knob to the correct value (HIGH=2 W, LOW=1 W). Press Menu to set.

8. Press F/A and then U/V to select memory to write to. Scroll using the knob to the desired memory. Press U/V again to write.

9. Hold U/V for two seconds to enter memory mode. Memories are available on the top side only.

10. Scroll to the channel you just wrote.

Lock/unlock radio

Hold Menu button for three seconds to lock/unlock.

Check repeater input frequency

There is no way to listen to the repeater input frequency except to program a separate memory for it.

Change power in the field

Press Menu and scroll to 07 - POWER to set power level. Press U/V to adjust and then scroll with the knob to the correct value (HIGH=2 W, LOW=1 W). Press Menu to set.

Adjust volume

Press Vol and use knob to adjust. Press Vol again to set.

Adjust squelch

Press Menu and scroll to 03 - SQL to set squelch. Press U/V to adjust and then scroll with the knob to the correct value (0–9). Press Menu to set.

Weird Modes

Radio shows RADIO

This radio has a broadcast FM receiver in it as well. If the radio displays RADIO on the first line of the LCD and a number on the second line, hold L/R for two seconds to get out of this mode.

Can't set offset / direction

This radio can display frequencies instead of channel names. If you have that enabled, it's difficult to tell if you're in channel (memory) mode or frequency (VFO) mode. One sure way is to try to program an offset or shift. If it doesn't "stick" after programming, that's likely because you're in channel mode. Switch to frequency mode in order to program.

Useful Information

The knob locks mechanically if you push it down. Pull it up to be able to rotate.

The battery can be removed without removing the screw in the back. Flip the switch on the bottom of the radio toward the front and the back panel will slide off.

The menu times out after about ten seconds. You have to be quick when setting values.

Although the radio advertises low power as 1 W, at least some versions output less than that (about 0.1 W) on VHF. There is a hardware fix that involves replacing an 0603 47 nH coil with a 150 nH coil.

Sometimes button presses don't register on this radio. It may be easier to program if you have the keyboard beep turned on. To do this, press Menu and scroll to 05 - BEEP to set beep. Press U/V to adjust and then scroll with the knob to the correct value (OFF, ON). Press Menu to set.

Factory reset

For full reset (loses all memories and settings) hold down Vol and hold Power for two seconds.

⚠ **WARNING:** In at least some configurations, the Baofeng UV-100, UV-200 and UV-3R may permit you to transmit on business or public safety frequencies. Make sure you are in-band when transmitting.

Radio Layout

Specs

Receivers Single receiver, dual watch (first to break squelch wins)
Receives 136–174 MHz and 400–480 MHz FM
Transmits 136–174 MHz @ 4 W FM and 400–480 MHz @ 4 W FM
Antenna connector SMA **M** on radio; needs SMA **F** antenna
Modes FM
Memory Channels 128
Power No DC input on radio
Model year 2012

Standard Tasks

Program frequency in the field

1. Decide which memory to use. You can't overwrite—you will have to delete first if that memory has data in it. Press Menu 2 8 Menu (DEL-CH) *XXX* where *XXX* is the channel (001–128). Then press Menu.

2. Press VFO/MR to go to VFO mode if you aren't there already.

3. Enter frequency. You may need to press Band to switch bands depending on your frequency.

4. Press Menu 1 3 Menu (T-CTCS) then scroll (using Up / Down) to the correct CTCSS tone frequency (or off), press Menu Exit.

5. Press Menu 2 5 Menu (SFT-D) and scroll (using Up / Down) to the correct repeater shift, press Menu Exit.

6. Press Menu 2 6 Menu (OFFSET) and enter or scroll (using Up / Down) to the correct repeater offset, then press Menu Exit.

7. Press Menu 2 Menu (TXP) and scroll (using Up / Down) to correct power level, then press Menu Exit.

8. Press Menu 2 7 Menu (MEM-CH) and enter channel to write *XXX* (001–128) then press Menu Exit.

9. Press VFO/MR to enter channel (memory) mode.

10. Scroll (using Up / Down) to the channel you just wrote.

Lock/unlock radio

Hold # for three seconds to lock/unlock.

Check repeater input frequency

There is no way to listen to the repeater input frequency except to program a separate memory for it.

Change power in the field

Press Menu 2 Menu (TXP) and scroll to correct power level (LOW is 1 W, HIGH is 4 W), press Menu Exit.

Adjust volume

Rotate power/volume knob to adjust volume.

Adjust squelch

Press Menu 0 Menu (SQL) then scroll (using Up / Down) to the correct squelch level, press Menu Exit.

Weird Modes

Can't leave channel (memory) mode

Some distributors ship these radios with VFO mode turned off. If that's the case, you will need to modify the programming with a computer / radio programming cable to turn VFO mode on. This cannot be changed from the front panel.

Can't set offset / direction

This radio can display frequencies instead of channel names. If you have that enabled, it's difficult to tell if you're in channel (memory) mode or frequency (VFO) mode. One sure way is to try to program an offset or shift. If it doesn't "stick" after programming, that's likely because you're in channel mode. Switch to frequency mode in order to program.

Useful Information

If you wait too long after pressing Menu, it will time out. You need to press quickly.

Hold down 3 while turning the radio on to see the firmware version. This radio comes in multiple colors.

Factory reset

Press Menu 4 0 Menu (RESET) then scroll (using Up / Down) to ALL, press Menu Exit.

VFO reset

Press Menu 4 0 Menu (RESET) then scroll (using Up / Down) to VFO, press Menu Exit.

⚠ **WARNING:** In at least some configurations, the Baofeng UV-5R may permit you to transmit on business or public safety frequencies. Make sure you are in-band when transmitting.

Baofeng UV-82

Radio Layout

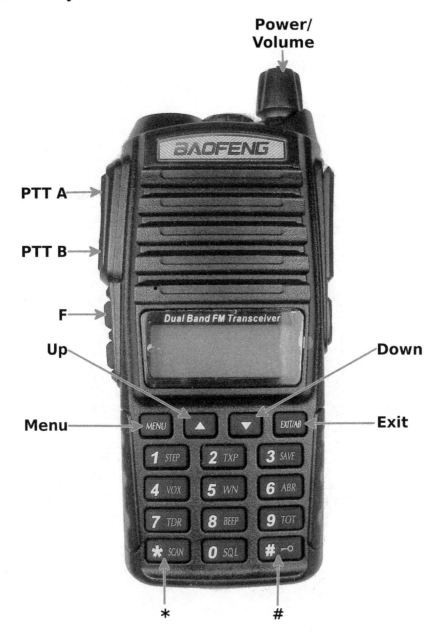

Specs

Receivers Single receiver, dual watch (first to break squelch wins)
Receives 136–174 MHz and 400–520 MHz FM
Transmits 136–174 MHz @ 5 W FM and 400–520 MHz @ 4 W FM
Antenna connector SMA **M** on radio; needs SMA **F** antenna
Modes FM
Memory Channels 128
Power No DC input on radio
Model year 2014

Standard Tasks

Program frequency in the field

1. Decide which memory to use. You can't overwrite—you will have to delete first if that memory has data in it. Press [Menu] [2] [8] [Menu] (DEL-CH) *nnn*, where *nnn* is the channel (001–128) to delete/program. (If the display shows CH-nnn the memory has data.) Then press [Menu].

2. If you're not already in VFO mode, turn off the radio, hold down [Menu] and turn the radio on to enter VFO mode.

3. Enter frequency.

4. Press [Menu] [1] [3] [Menu] (T-CTCS) then enter the correct PL tone frequency (1622 for 162.2, and 0 for OFF), press [Menu] [Exit].

5. Press [Menu] [2] [5] [Menu] (SFT-D) and scroll using [Up] and [Down] to the correct repeater shift, press [Menu] [Exit].

6. Press [Menu] [2] [6] [Menu] (OFFSET) and scroll to the correct repeater offset, then press [Menu] [Exit].

7. Press [Menu] [2] [Menu] (TXP) and scroll to correct power level, then press [Menu] [Exit].

8. Press [Menu] [2] [7] [Menu] (MEM-CH) and enter channel to write *XXX* (001–128) then press [Menu] [Exit].

9. Turn off the radio, hold down [Menu] then turn it back on to enter channel (memory) mode.

10. Scroll to the channel you just wrote.

Lock/unlock radio

Hold [#] for two seconds to lock/unlock.

Check repeater input frequency

Press [*] (display shows R). Press [*] again to revert to normal.

Change power in the field

Press ⎡ # ⎤ to switch between low power (1 W; display shows (L) and high power (5 W on 2m, 4 W on 70cm; display shows nothing).

Adjust volume

Rotate power/volume knob to adjust volume.

Adjust squelch

Press ⎡Menu⎤⎡ 0 ⎤⎡Menu⎤ (SQL) then scroll to the correct squelch level, press ⎡Menu⎤⎡Exit⎤.

Weird Modes

Can't leave channel (memory) mode

The Baofeng UV-82C is FCC Part 90 accepted, and must be unlocked using a computer / radio programming cable to turn VFO mode on. This cannot be changed from the front panel. The Baofeng UV-82 is not Part 90 accepted. Both these radios can transmit on business frequencies. Be careful to make sure that you are in-band when transmitting.

Radio makes alarm noise

If you hold down ⎡ F ⎤ for two seconds, the radio enters Alarm Mode. **This transmits an alarm on the currently selected frequency.** Turn the radio off and back on to clear it.

Can't set offset / direction

This radio can display frequencies instead of channel names. If you have that enabled, it's difficult to tell if you're in channel (memory) mode or frequency (VFO) mode. One sure way is to try to program an offset or shift. If it doesn't "stick" after programming, that's likely because you're in channel mode. Switch to frequency mode in order to program.

Useful Information

The Baofeng UV-82 has two PTT buttons. Normally ⎡PTT A⎤ acts as PTT for the first frequency on the display, while ⎡PTT B⎤ acts as PTT for the second frequency. On the UV-82C, ⎡PTT A⎤ can be disabled in software.

If you wait too long after pressing ⎡Menu⎤, it will time out. You need to press quickly.

Hold down the ⎡Menu⎤ button while turning the radio on to switch from VFO mode to Memory (Channel) mode or vice versa.

Factory reset

Press [Menu] [4] [1] [Menu] (RESET) then scroll (using [Up] / [Down]) to
ALL, press [Menu] [Exit]. After reset, the 2m offset is set to 1.600 MHz. Use
menu 26 to change it.

VFO reset

Press [Menu] [4] [1] [Menu] (RESET) then scroll (using [Up] / [Down]) to
VFO, press [Menu] [Exit].

⚠ **WARNING:** In at least some configurations, the Baofeng UV-82 may permit you to transmit on business or public safety frequencies. Make sure you are in-band when transmitting.

BridgeCom BCH-220

Radio Layout

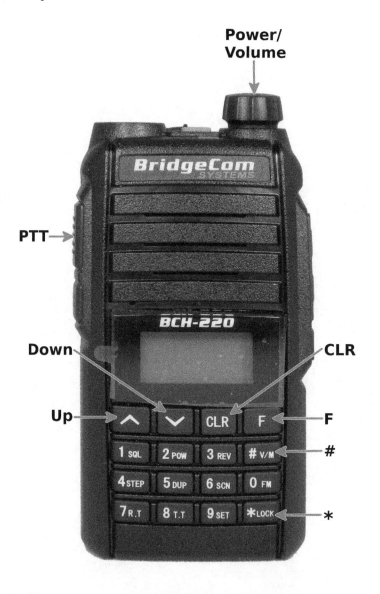

Specs

Receivers Single receiver, dual watch (first to break squelch wins)
Receives 219–260 MHz FM
Transmits 222.0–224.995 MHz @ 5 W FM
Antenna connector SMA **M** on radio; needs SMA **F** antenna
Modes FM
Memory Channels 199
Power No DC input on radio, 7.4 V DC per spec
Model year 2015

Standard Tasks

Program frequency in the field

1. If you're not already in VFO mode, press [#] (V/M) to go to VFO mode. (In memory mode, you'll see the channel number in the upper left; you won't see this in VFO mode.)

2. To set the frequency, enter using the numeric keypad (223500 for 223.500 MHz).

3. If you need to set the offset to something other than the standard offset, press [F] [9] [9] [Up] to select menu 10, OFFSET. Press [F] then enter frequency using the numeric keypad (001600 for 1.6 MHz). Press [CLR] to exit.

4. To set the duplex, press [F] [5] to cycle through no offset, positive offset and negative offset. There is no indication on the front panel about the offset; you will need to watch the frequency on the display when you press [PTT] to see which way you have set. This **does transmit**.

5. To set tone, press [F] [8] (T.T). Press [*] to cycle through OFF, CTCSS tones, and DCS tones (begin with D). Use [Up] / [Down] to select the one you want. (You can enter CTCSS tones with the keypad to get things set a little faster; 0670 for 67.0 Hz.) Then press [F]. This sets transmit tone; to set receive tone you can use [F] [7] (R.T).

6. To set transmit power, press [F] [2] (POW) then use [Up] / [Down] to choose POW HI (high power, 5 W) or POW LOW (low power, 2 W). Press [F] to set.

7. To write the memory, press Press [F] [*]. Then use [Up] / [Down] or enter the memory to write directly from the keypad (012 is memory 12). Press [F] to save.

8. To switch to memory mode, press [#] (V/M).

9. In memory mode, use [Up] / [Down] or enter a memory number to go to the memory you just wrote.

Lock/unlock radio

Hold [*] (LOCK) for two seconds to lock/unlock. Lock icon will show when radio is locked.

Check repeater input frequency

Press [F] [3] (REV) to switch to monitoring input frequency. Press [F] [3] (REV) again to revert to normal.

Change power in the field

Press [F] [2] (POW) then use [Up] / [Down] to choose POW HI (high power, 5 W) or POW LOW (low power, 2 W). Press [F] to set.

Adjust volume

Rotate power/volume knob to adjust volume.

Adjust squelch

Press [F] [1] (SQL) to change squelch, then use [Up] / [Down] to select squelch level (0–9). Press [F] to set.

Weird Modes

Display shows OFF when PTT is pressed

This means the radio is trying to transmit out of the limits of the transmit band. Adjust the offset and offset direction.

Radio makes alarm noise

This radio has an alarm mode which can be assigned to a button. If you hold down the button that is assigned to the alarm for two seconds, the radio enters Alarm Mode. **This transmits an alarm on the currently selected frequency**. Turn the radio off to clear it.

Radio displays FM broadcast frequencies

This radio claims to be able to receive FM broadcast frequencies. Press [F] [0] (FM) to switch back to 220 MHz.

Radio will not enter VFO mode

This radio has a channel-only mode. To enter or exit it, hold down [#] while turning the radio on.

Radio locks itself after an interval

The radio has a feature to lock out the keypad after a certain amount of time. Press F 9 (SET), then press F . Use Up / Down to select menu 02, LOCK.KEY. Press F and use Up / Down to select from MANU (no auto-lock), AT 5 (lock after five seconds), AT 10 (lock after ten seconds), AT 20 (lock after twenty seconds), AT 30 (lock after thirty seconds), ATS 5 (lock after five seconds and preserve on power off), ATS 10 (lock after ten seconds and preserve on power off), ATS 20 (lock after twenty seconds and preserve on power off), and ATS 30 (lock after thirty seconds and preserve on power off). Press F to set and CLR to exit.

Offset cannot be changed

You must be in VFO mode to change the offset. Press # to switch modes.

Useful Information

The top and side buttons are programmable. By default the side is a monitor button and the top activates alarm mode.

You can enter the set menu by pressing F 9 (SET), and navigate it using Up / Down. If you wait too long after entering set mode, it will time out. You need to press quickly. In set mode, press F to change the value of the currently selected parameter, and press F to set it. Use CLR to exit after setting.

Factory reset

To reset the radio, enter set mode and use Down until you see menu 18, RESET. Press F . Use Down to choose FULL (full factory reset). For a full reset, you will be prompted for a password—the default is 000000. Enter the password, then press F to reset. Note that after reset, offset reverts to 0.000 MHz and will need to be changed.

Settings reset

To reset the radio, enter set mode and use Down until you see menu 18, RESET. Press F . Use Down to choose either VFO (this will reset everything except memories). Then press F to reset. Note that after reset, offset reverts to 0.000 MHz and will need to be changed.

Harris Unity XG-100P

Radio Layout

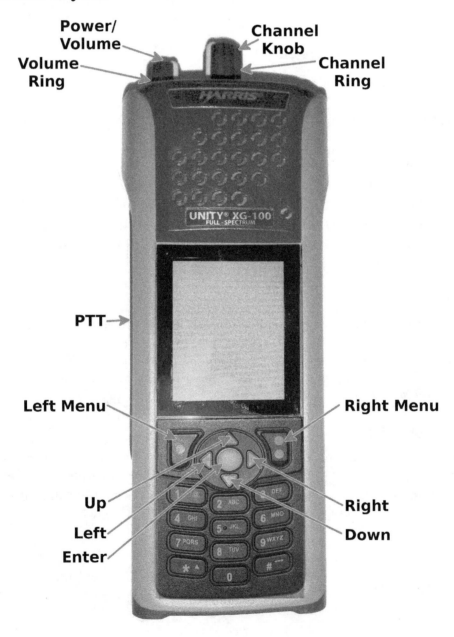

Specs

Receivers Single receiver
Receives 136–174 MHz FM, 380–520 MHz FM, 762–870 MHz FM
Transmits 136–174 MHz @ 6 W FM, 380–520 MHz @ 5 W FM, 762–870
   MHz @ 3 W FM
Antenna connector SMA F on radio with special recessed connector; uses
   130–870 MHz multiband antenna with SMA M connector
Modes FM, P25
Memory Channels 1000 per system, 512 systems
Power No DC input on radio; 7.5 V DC nominal, VC4000 charger input is
   11–16 V DC
Model year 2010

Standard Tasks

Program frequency in the field

1. This radio has an unlock code that is required to enable front panel
   programming. If you don't know the unlock code, you're out of luck.

2. Select the channel you want to program with the channel selector
   knob (and often the channel selector ring). You will overwrite the
   selected channel. You must use an existing analog channel; you can't
   overwrite a digital channel.

3. Make sure you're on the main screen, then press [Right Menu] (while
   display shows CH INFO).

4. Press [Right Menu] (while display shows EDIT CHAN).

5. Enter the front panel programming unlock password by pressing the
   appropriate numbers on the keypad, then [Enter].

6. Use [Up] and [Down] to scroll to RX FREQUENCY. Press [Enter]. Enter
   the frequency using the keypad (144390 for 144.390 MHz). Press
   [Enter].

7. Use [Up] and [Down] to scroll to TX FREQUENCY. Press [Enter]. Enter
   the frequency using the keypad (144390 for 144.390 MHz). Press
   [Enter].

8. To set transmit power, use [Up] and [Down] to scroll to TX POWER.
   Press [Enter] to switch between LOW and HIGH. (Note that the values
   for low and high power must be set by programming software.)

9. To set receive tone, use [Up] and [Down] to scroll to menu entry
   RX CHAN GUARD. Press [Enter] to bring up the menu to choose from
   NOISE (no tone), CTCSS (CTCSS tone) and CDCSS (DCS tones). Press
   [Enter] to select. Next scroll to RX TONE, press [Enter], scroll to the
   appropriate tone and press [Enter].

10. To set transmit tone, use [Up] and [Down] to scroll to menu entry
    TX CHAN GUARD. Press [Enter] to bring up the menu to choose from
    NOISE (no tone), CTCSS (CTCSS tone) and CDCSS (DCS tones). Press
    [Enter] to select. Next scroll to TX TONE, press [Enter], scroll to the
    appropriate tone and press [Enter].

11. Press [Left Menu] twice to get back to the main screen.

Lock/unlock radio

The way lock is enabled can be configured through software, but many
users have the volume ring configured as a lock button. Rotate it to ⊘ to
lock and ◯ to unlock. An alternate method if keypad lock is turned on is
to press [Left], [Right], [Up], [Down].

Check repeater input frequency

You can't switch to the repeater input frequency in the field. You will have
to program a separate memory for this.

Change power in the field

Follow the process for editing a channel. Using [Up] and [Down], select
TX POWER. Press [Enter] to switch between HIGH and LOW power. Depending
on programming, the channel selector ring can also be configured to switch
to high power. Power levels for high and low power cannot be configured
in the field, but high power is usually 6 W / 5 W / 3 W depending on
frequency.

Adjust volume

Turn the power/volume knob to adjust volume.

Adjust squelch

Squelch cannot be set in the field. Squelch level must be programmed into
the radio via PC.

Weird Modes

Buttons don't work

Many of this radio's buttons and switches are programmable using a PC
programmer. Depending on how the radio was configured, the buttons
may or may not work, and may do different things.

Radio has open mic, flashing lights and/or sounds

The radio has an emergency mode. This transmits a signal on P25, but
on analog channels it just opens the mic. The radio will also display

EMERGENCY. Depending on the emergency mode programmed, there may also be exciting lights and sounds. Turn the radio off and on again to exit.

Useful Information

The volume ring may be programmed for one of many functions. They include encryption (⊘ is encrypted), TX disable (⊘ is no transmit), talkaround (⊘ is talkaround on), lock (⊘ is locked) and scan (⊘ is scanning).

This radio has multiple sets of memories (called "mission profiles"). If you make a change and then load a different mission profile, your change will be overwritten.

⚠ **WARNING:** This radio uses a recessed SMA-style connector and a multi-band antenna. Some SMA antennas may not fit.

SMA connector on antenna   SMA jack on radio

⚠ **WARNING:** This radio is normally used in trunked systems. It will transmit its ID on such systems even if you don't push the PTT. Don't turn the radio on without an appropriate antenna when it's on a trunked channel.

No reset procedure

This radio doesn't have a reset, but you can reload an existing mission plan. To do this from the main screen, press 7 and select a plan using Up and Down. When you have the plan selected, press Right Menu (when the display shows OPTIONS). Select the menu entry ACTIVATE PLAN using Up and Down, then press Enter again to select. You may be asked for a PIN if the mission plan has one set.

⚠ **WARNING:** In at least some configurations, the Harris Unity XG-100p may permit you to transmit on business or public safety frequencies. Make sure you are in-band when transmitting.

Icom IC-P2AT

Radio Layout

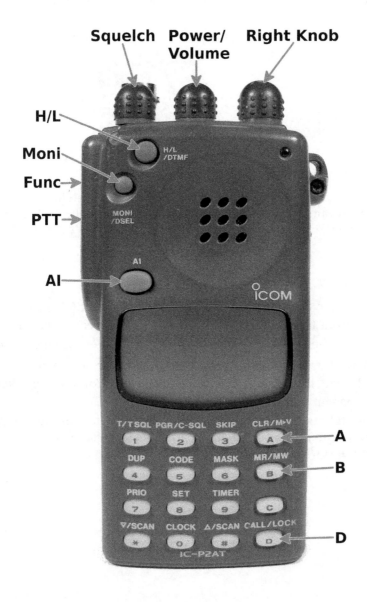

Specs

Receivers Single receiver
Receives 138–174 MHz FM
Transmits 140–150 MHz @ 5 W FM
Antenna connector BNC F on radio; needs BNC M antenna
Modes FM
Memory Channels 100
Power 6–16 V DC per spec, 13.8 V DC nominal, 3.5mm OD, 1.3mm ID
    barrel style plug, center positive
Model year 1991

Standard Tasks

Program frequency in the field

1. To select memory to write to, press B (MR) to enter memory mode.
   Hold down Func then scroll with right knob to select a memory.

2. Press A (CLR) to go to VFO mode.

3. Enter last four digits of frequency on numeric keypad (so 144.390 is
   4390). Note that if you're outside of transmit range, this will change
   only the last four digits of the frequency (e.g. at 160.0, pressing 4390
   will change to 164.390). To get back into range quickly, hold down
   Func and press Moni until there is a bar under the third digit of the
   frequency. Continue to hold down Func, then use the right knob to
   adjust frequency appropriately.

4. For repeater tone, hold Func and press 8 (SET). Press * (Down)
   / # (Up) until you see TO (repeater tone). Scroll to the correct tone
   using the right knob. Press A (CLR) to select. Hold Func and press
   1 (T/T SQL) to activate tone (cycles through Tone, Tone Squelch
   with Beep, Tone Squelch, None). T is shown when tone is active.

5. Select repeater shift by holding Func and pressing 4 (DUP) to select from -DUP (-) / DUP (+) /none.

6. For repeater offset, hold Func and press 8 (SET). Press * (Down)
   / # (Up) until you see OW (offset). Use the right knob to scroll to the
   correct offset. Press A (CLR) to set. .60 is 600 MHz.

7. For power level, hold down H/L and use the right knob to select from
   high (5 W) or one of the three low levels (3.5 W, 1.5 W and 0.5 W).
   The S-meter indicates the level. Just pressing H/L switches between
   high and the last low level set.

8. To write memory, hold Func and hold B (MW) for two seconds.

9. Press B to go back to memory mode, then use the right knob to
   scroll to the channel you just wrote.

Lock/unlock radio

Hold Func then press D (Lock) button to lock/unlock.
    There's also a PTT lock. See "Weird Modes" for more details.

Check repeater input frequency

There is no way to listen to the repeater input frequency except to program
a separate memory for it.

Change power in the field

Hold down H/L and use the right knob to select from high (5 W) or one
of the three low levels (3.5 W, 1.5 W and 0.5 W). The S-meter indicates the
level. Just pressing H/L switches between high and the last low level set.

Adjust volume

Rotate center power/volume knob.

Adjust squelch

Rotate left squelch knob.

Weird Modes

PTT doesn't transmit (PTT lock)

The IC-P2AT has a PTT lock function. To disable it, press A to enter VFO
mode. Then hold Func and press 8 (SET). Press * (Down) / # (Up)
until you see PT. Use the right knob to scroll to P (allow PTT) or PL (prevent
PTT). Press A (CLR) to set.

Not all functions are available

The IC-P2AT has a simplification feature that prevents users from entering
some functions. When all features are enabled, you will see five stars at
the top of the display. If you see fewer than five, you may not be able to
perform some functions. To enable all functions, turn the radio off, then
hold down H/L and AI and turn the radio on. You will be prompted with
StAr. Use the right knob to adjust to five stars, then press PTT to set
(does not transmit).

Useful Information

A tone module (either UT-50 or UT-51) must be installed to use CTCSS
tones. The UT-50 does not provide 97.4 Hz.

Factory reset

To reset the CPU, hold down Func and A while turning the power on.
This also sets all memories and setting to default.

Icom IC-T70A

Radio Layout

Specs

Receivers Single receiver, dual watch (first to break squelch wins)
Receives 138–174 MHz FM, 400–479 MHz FM
Transmits 144–148 MHz @ 5 W FM, 420–450 MHz FM
Antenna connector SMA F on radio; needs SMA M antenna
Modes FM
Memory Channels 250
Power 10–16 V DC, 3.5mm OD, 1.3mm ID barrel style plug, center positive
Model year 2010

Standard Tasks

Program frequency in the field

1. Press C until you are in VFO mode. (When in VFO mode, you won't see C or WX on the main LCD, and the MR icon will not be visible.)

2. Press B to select band.

3. Enter the receive frequency on the numeric keypad without decimal (144.390 is 144390).

4. For repeater tone, press A then rotate knob until you see R tONE. Rotate the ring to select the tone value. (You can also set tone squelch C tONE and DCS code codE.)

5. For repeater offset, rotate knob until you see OFFSEt. Rotate ring to select the offset value.

6. Press C to leave set mode.

7. To set tone mode, repeatedly hold for one second and then release 0 to cycle through T (tone), ((.)) T SQL (tone squelch with beep), T SQL (tone squelch), ((.))DTCS (DCS with beep), DTCS (DCS), T SQL-R (squelch if a particular tone *is* received), DTCS -R (squelch if a particular code *is* received), and blank (no tone/code).

8. Select repeater shift by repeatedly holding for one second and then releasing * to cycle through DUP - (-), DUP (+) and blank (simplex).

9. For power level, press C repeatedly to cycle through high (5 W), medium (2.5 W, radio shows M) and low (0.5 W, radio shows L).

10. To write memory, hold D for one second and release. Use knob to scroll to a memory (0 through 249. Stay away from channels with A or b in them; those are scan edge channels.) Hold D for one second and release to write.

11. Press D to go back to memory mode.

Lock/unlock radio

Hold [A] for one second to lock/unlock.
    There's also a PTT lock. See "Weird Modes" for more details.

Check repeater input frequency

Hold [B]. Radio will enter monitor mode and switch to reverse frequency.
Release [B] to go back to normal.

Change power in the field

Press [C] repeatedly to cycle through high (5 W), medium (2.5 W, radio
shows M) and low (0.5 W, radio shows L).

Adjust volume

Rotate ring; volume is from 0–24.

Adjust squelch

Hold down [B] and rotate knob. Settings are OPEN, Auto, then LEVEL1
through LEVEL9.

Weird Modes

Radio displays Hot

The radio has protection from overheating. It will first fold power back to
medium from high, and then prevent transmission entirely. Cool the radio
down.

PTT doesn't transmit (PTT lock)

The IC-T70A has a PTT lock function. To disable it, hold down press [A]
while turning on the radio. Rotate the knob until you see Ptt Lk. Rotate
the ring until you see OFF PtL. Press [Power]. To turn it back on, repeat
the same steps using the ring to select ON PtL.

Not all frequencies are available in memory mode

The IC-T70A can have memory banks configured. If it does, you can turn
them off. While in memory mode, press [B]. Next rotate the knob until
you don't see bAnk. Then press [B] again.

Useful Information

Rotating the ring on this radio can sometimes cause the knob to rotate as
well, and vice versa. Hold one down carefully while adjusting the other.

The IC-T70A speaker mic has an input impedance of 2.2 kΩ. Some third-party speaker mics may not be properly matched.

The display will show two different kinds of WX. There's a WX icon which is visible when weather alert is turned on. If you're actually on the weather band, you'll see WX-01 through WX-10 on the main LCD display.

Factory reset

To reset everything including memories, hold down $\boxed{A}$, $\boxed{B}$ and $\boxed{C}$ while turning the radio on. The radio will reset immediately with no prompt.

Settings reset

To reset most settings but leave memories intact (partial reset), hold down $\boxed{A}$ while turning the radio on. The radio will reset immediately with no prompt.

Icom IC-V8

Radio Layout

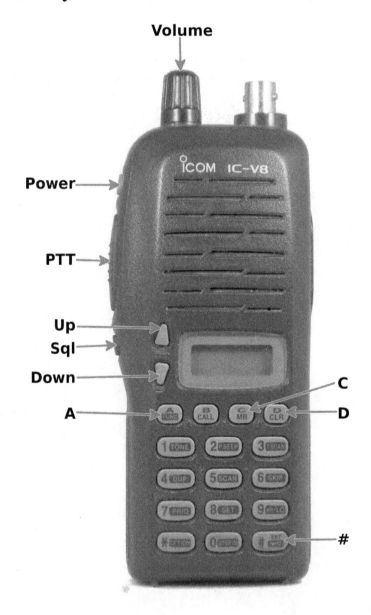

Specs

Receivers Single receiver
Receives 136–174 MHz FM
Transmits 144–148 MHz @ 5.5 W FM
Antenna connector BNC F on radio; needs BNC M antenna
Modes FM
Memory Channels 100
Power No DC input on radio; 6–10.3 V DC per spec
Model year 2001

Standard Tasks

Program frequency in the field

1. Press D (CLR) to go to VFO mode if you aren't there already. (Screen will show MR if you aren't.)

2. Enter frequency on numeric keypad (all digits, so 144.390 is 144390).

3. For repeater tone, press A (FUNC) 8 (SET) and push Up / Down until you see rt (repeater tone). Scroll to the correct tone using the volume knob. Press # (ENT/Lock) to select. Press A (FUNC) 1 (TONE) to activate tone. (Musical note is shown when tone is active. Other options are a speaker symbol (tone squelch), D (DCS) and none.)

4. This radio has automatic repeater shift. If it gets it wrong, repeatedly press A (FUNC) 4 (DUP) to select from -/+/none.

5. For repeater offset, press A (FUNC) 8 (SET) and push Up / Down until you see ± (offset). Scroll to the correct tone using the volume knob. Press # (ENT/Lock) to select. 0.60 is 600 MHz.

6. For power level, press A (FUNC) then 9 (HI/LOW) to select between 5.5 W (shows nothing on right) and 0.5 W (shows L on right).

7. To write to memory, press A (FUNC) C (MR). MR will flash. Use Up / Down to select desired channel. Press A (FUNC) then hold C (MR) for one second to write memory.

8. Press C (MR) to enter channel (memory) mode.

9. Scroll to the channel you just wrote. Note that D (CLR) goes back to VFO mode; after pressing it you'll need to press C (MR) again.

Lock/unlock radio

Press A (FUNC) then hold # (ENT/Lock) button for one second to lock/unlock.

Check repeater input frequency

Press $\boxed{D}$ (CLR) if you're not already in VFO mode, then $\boxed{A}$ (FUNC) $\boxed{8}$ (SET) and push $\boxed{Up}$ / $\boxed{Down}$ until you see REV.OF (reverse mode off). Use the volume knob to change to REV.ON, then press $\boxed{\#}$ (ENT/Lock) to select. In reverse mode the + or - symbol will blink. Repeat procedure to switch back to REV.OF.

Change power in the field

Press $\boxed{A}$ (FUNC) then $\boxed{9}$ (HI/LOW) to select between 5.5 W (shows nothing on right) and 0.5 W (shows L on right).

Adjust volume

Turn the volume knob to adjust volume.

Adjust squelch

Hold down $\boxed{Sql}$ button and push $\boxed{Up}$ / $\boxed{Down}$.

Weird Modes

Volume knob changes frequency rather than volume

The IC-V8 has an automobile mode which switches the functions of the volume knob and the $\boxed{Up}$ / $\boxed{Down}$ buttons. If you're in it, turn the radio off. Then hold down both $\boxed{Up}$ and $\boxed{Down}$ while turning the radio on. Scroll with the $\boxed{Up}$ / $\boxed{Down}$ buttons get to toP.dl (knob is for tuning). Then use the volume knob to change it to toP.VO (knob is for volume). Press $\boxed{\#}$ (ENT/Lock) to select.

Radio doesn't transmit

The IC-V8 can be set to prevent transmission. Use $\boxed{D}$ (CLR) if you're not already in VFO mode, then $\boxed{A}$ (FUNC) $\boxed{8}$ (SET) and scroll with $\boxed{Up}$ / $\boxed{Down}$ to tX. Use volume to adjust between tX.ON (allows transmit) and tX.OF (prevents transmit). Press $\boxed{\#}$ (ENT/Lock) to set.

Function key works strangely

The behavior of the $\boxed{A}$ (FUNC) key can be changed. Use $\boxed{D}$ (CLR) if you're not already in VFO mode, then $\boxed{A}$ (FUNC) $\boxed{8}$ (SET) to change. F0.AT is the default, meaning function mode disappears zero seconds after using a function. It can also be set to F1.AT, F2.AT, F3.AT (function mode disappears one, two, three seconds after using a function) or F.m (you have to hit $\boxed{A}$ (FUNC) to turn off function mode).

Can't get out of memory mode

The IC-V8 has a channel mode which prevents direct frequency entry. In this mode, channels will display as CH 27. To get out of it, turn the radio off. Then hold down both [Up] and [Down] while turning the radio on. Scroll with the [Up] / [Down] buttons get to dsp.CH (display channel mode). Use the volume knob to change it to dsp.FR (frequency) or dsp.Nm (name). Press [#] (ENT/Lock) to select.

Useful Information

⚠ **WARNING:** According to an article from Icom (see the Icom website at http://www.icom.co.jp/world/notice/) the IC-V8 has been counterfeited, and the counterfeit is being sold at online auction sites, among other places. The counterfeit radios can be identified by a number of ways, the most obvious being that the [2] reads "VOX" instead of "P.BEEP," and the speaker mic jacks have white labels instead of no labels. The menus for these counterfeit devices are different from the Icom menus.

Tone squelch can use a different frequency than CTCSS tone. Use the same procedure to adjust tone squelch as you do for tone, but scroll to Ct for tone squelch and dt for DCS.

Factory reset

To reset the CPU, hold down [D] (CLR) and [SQL] while pushing the [Power] button. This also clears all memories and settings.

Icom IC-V80 HD

Radio Layout

Specs

Receivers Single receiver with priority channel option
Receives 136–174 MHz FM
Transmits 144–148 MHz @ 5.5 W FM
Antenna connector BNC F on radio; needs BNC M antenna
Modes FM
Memory Channels 200
Power No DC input on radio; 7.2–9 V DC based on available batteries
Model year 2010

Standard Tasks

Program frequency in the field

1. Press VFO/MR/CALL to go to VFO mode if you aren't there already. (Screen will show MR, CO or WX if you aren't.)

2. Enter frequency on numeric keypad (all digits, so 144.390 is 144390).

3. For repeater tone, press * (FUNC) 8 (SET) and push Up / Down until you see rt (repeater tone). Scroll to the correct tone using the knob. Press # (ENT/Lock) to set.

4. Press * (FUNC) 1 (TONE) to activate tone. (Musical note is shown when tone is active. Other options are speaker symbol (tone squelch), speaker with lines coming out (tone squelch with "pocket beep" paging), D with a musical note (DCS transmit, CTCSS squelch), D with lines (DCS squelch with "pocket beep"), D (DCS) and none.)

5. This radio has automatic repeater shift. If it gets it wrong, repeatedly press * (FUNC) 4 (DUP) to select from -/+/none.

6. If you need to change repeater offset frequency, press * (FUNC) 8 (SET) to adjust. Use Up/Down to get to ±. Use the knob to adjust the value, and press # (ENT/Lock) to set.

7. To adjust power, press * (FUNC) 9 (H/M/L) to select from 5.5 W (shows nothing on screen), 2.5 W (shows M on screen) and 0.5 W (shows L on screen).

8. To write to memory, press * (FUNC) then hold VFO/MR/CALL. Use Up / Down to set memory, then hold VFO/MR/CALL to write.

9. Press VFO/MR/CALL to enter memory mode.

10. Scroll to the channel you just wrote.

Lock/unlock radio

Press * (FUNC) then hold # (ENT/Lock) button for one second to lock/unlock.

Check repeater input frequency

Press [*] (FUNC) [8] (SET) and push [Up] / [Down] until you see REV.OF (reverse mode off). Use the knob to change to REV.ON, then press [#] (ENT/Lock) to select. In reverse mode the + or - symbol will blink. Repeat procedure to switch back to REV.OF. You can also hold [Moni] to temporarily check the input frequency of a repeater memory.

Change power in the field

To adjust power, press [*] (FUNC) [9] (H/M/L) to select from 5.5 W (shows nothing on screen), 2.5 W (shows M on screen) and 0.5 W (shows L on screen).

Adjust volume

Turn the knob to adjust volume.

Adjust squelch

Hold down [Moni] button and push [Up] / [Down] to go from level 0–10.

Weird Modes

Knob changes frequency rather than volume

This radio can be configured so the knob changes frequency instead of volume. To adjust, turn the radio off. Then turn it on while holding [Up] and [Down]. Press [Up] / [Down] to scroll to tOP.. Use the knob to select from tOP.VO (volume) or tOP.dI (frequency). Press [#] (ENT/Lock) to select.

Radio doesn't transmit

This radio has a transmit inhibit function which can be stored with individual memories. To disable, press [*] (FUNC) [8] (SET) to enter set mode and and push [Up] / [Down] until you see tX OF. Scroll using the knob to tX ON. Press [#] (ENT/Lock) to set.

Can't get out of memory mode

This radio can be configured to use a memory-only mode. To exit, turn the radio off. Then turn it on while holding [Up] and [Down]. Press [Up] / [Down] to scroll to dSP.Ch. Use the knob to select either dSP.FR (display frequency) or dSP.Nm (display name). Press [#] (ENT/Lock) to select.

Useful Information

Hold [Power] for one second to turn on/off.

Tone squelch can use a different frequency than CTCSS tone. Use the same procedure to adjust tone squelch as you do for tone, but scroll to `Ct` for tone squelch and `dt` for DCS.

Factory reset

To reset everything, turn the radio off. Then hold down VFO/MR/CALL and Moni and while turning the radio on. The display will show `CLEAR` and the radio will be completely cleared.

Settings reset

To reset settings, turn the radio off. Hold down VFO/MR/CALL while turning the radio on. Nothing obvious will happen but the settings will be reset.

Icom IC-W32A

Radio Layout

Specs

Receivers Two independent receivers, simultaneous receive
Receives 118–135.995 MHz AM, 136–174 MHz FM, 440–470 MHz FM
Transmits 144–148 MHz @ 5 W FM, 440–450 MHz @ 5 W FM
Antenna connector BNC F on radio; needs BNC M antenna
Modes FM
Memory Channels 100 VHF, 100 UHF
Power 4.5–16 V DC per spec, 13.5 V DC nominal, 3.5mm OD/1.3mm
ID center positive (on battery—different batteries may have different
connectors)
Model year 1996

Standard Tasks

Program frequency in the field

1. Press [MAIN] to select appropriate band.

2. Press [VFO] to go to VFO mode.

3. Enter the frequency on numeric keypad (no decimal required, so 144.390 is 144390).

4. Hold [H/L] for two seconds to enter set mode.

5. For repeater tone, press [H/L] or [TONE] until you see RT. Use the appropriate knob (left or right) to select the desired tone. You can also use [H/L] and [TONE] to move to CT to set a different tone for tone squelch if desired.

6. For repeater offset, press [H/L] or [TONE] until you see OW. Use the appropriate knob to adjust offset (global for selected band).

7. Press [VFO] to exit set mode.

8. Select repeater shift by holding [TONE] for two seconds. Repeat to cycle through -DUP (negative), DUP (positive) and blank (no shift).

9. Press [TONE] repeatedly to enable tone encoder. Display will cycle through T (tone on transmit, no tone on receive), T SQL(.) (tone page), T SQL (tone on transmit, tone on receive).

10. For power level, press [H/L] to select between high power (5 w) and low power (0.5 W, display shows LOW).

11. To enter memory writing mode press [S.MW].

12. Use the appropriate knob to select the correct memory.

13. Hold [S.MW] for two seconds to write.

14. Press [MR] to go back to memory mode, then use knob to select memory.

Lock/unlock radio

Hold CALL for two seconds to lock/unlock. Lock just prevents frequency change.

Check repeater input frequency

There is no way to listen to the repeater input frequency except to program a separate memory for it.

Change power in the field

To change power level, press H/L to select between high power (5 W) and low power (0.5 W, display shows LOW).

Adjust volume

Rotate left volume ring for left band volume; rotate right volume ring for right band volume.

Adjust squelch

Hold down SQL while rotating left knob or right knob. Values are AT (auto squelch), OPN (open) and SQ1 through SQ8.

Weird Modes

Radio shows CH:01 or 118:00

The radio can tune to NOAA weather channels (CH:01 through CH:10) and the air band (118–135.995 MHz). Press BAND repeatedly until you see the 2m and 70cm bands.

Radio shows HCH:nn or LCH:nn

The radio can display either channel names or channel numbers. Push M.N to switch between them.

Display is hard to read

The LCD contrast is adjustable. To change, hold H/L while turning the radio on. Then press TONE nine times. If you can see the display, it will show LC. Rotate the right knob to select contrast. Turn the radio off to save.

Display shows LOW V or OVER V when turned on

LOW V means the unit has less than 4.5 V available. OVER V means the device is getting more than 16 V.

Useful Information

The buttons H/L and TONE are sometimes used to scroll up/down.

If you write into a memory that previously had a channel name, that name will be retained. This may be confusing.

The device has an initial setup menu that is entered by holding H/L while turning on the device. From there, you can adjust MS (handheld mic mode), AO (auto power off), LI (backlight), BE (keyboard beep), AR (auto repeater shift), PS (power saver), VO (battery voltage display), DT (DTMF speed), LC (LCD contrast) and CB (crossband repeat mode). Use H/L / TONE to scroll through them, and set the value with either knob. Turn the power off again to save the values.

This radio has a built-in manual. Hold down L/G while pressing any button to learn more about that button. Press any button to exit.

This radio can crossband repeat. Set up is done using the initial setup menu. Modes are SEMI CB (mute sub-band when main band is transmitting) and FULL CB (two-way). To turn on crossband repeat: set the main and sub band frequencies, then lock the radio by holding CALL. Then turn the radio off. Finally hold down MAIN, BAND and SQL while turning the radio on. The lock icon will flash. Unlock (by holding CALL) to turn off crossband repeat.

The radio can be put in an extended receive mode by holding down MAIN, BAND, CALL and SQL while turning the radio on.

Factory reset

To reset everything (including memories), hold down SQL, VFO and MR while turning the power on.

Settings reset

To reset the CPU and settings only (but not memories), turn the radio on while holding down VFO.

Icom ID-51A

Radio Layout

Specs

Receivers Two independent receivers, simultaneous receive
Receives 108–174 MHz FM, DSTAR, AM, 390–479 MHz FM, DSTAR, AM
Transmits 144–148 MHz @ 5 W FM, DSTAR, 430–45- MHz @ 5 W FM, DSTAR
Antenna connector SMA F on radio; needs SMA M antenna
Modes FM, DSTAR
Memory Channels 500
Power 12–16 V DC, 3.5mm OD/1.3mm ID center positive
Model year 2012

Standard Tasks

Program frequency in the field

1. Press [V/MHz] to go to VFO mode if you aren't already there.

2. Select the band. Press [QUICK] to open the menu, use [Up] / [Down] to select Band Select, then press [Enter]. Use [Up] / [Down] to choose the band you want, then press [Enter].

3. Set the step size if needed. Press [QUICK] then use [Up] / [Down] to select TS. Press [Enter]. Use [Up] / [Down] to select the step size you want, then press [Enter].

4. Set the mode. Press [MODE] to cycle through FM, FM-N, DV (DSTAR) and AM (no transmit on AM).

5. Use knob to select frequency. Press [V/MHz] to cycle through 1 MHz adjustments, 10 MHz adjustments or normal step size.

6. To set repeater shift, press [MENU]. Use [Up] / [Down] to navigate to DUP/TONE. Press [Enter]. Use [Up] / [Down] to select DUP. Press [Enter]. Use [Up] / [Down] to select DUP- (negative offset), DUP+ (positive offset) or OFF (simplex). Press [Enter] to set. Press [MENU] to exit the menu.

7. To set the offset, press [QUICK] then use [Up] / [Down] to select Offset Freq. Press [Enter]. Use the knob to adjust offset. Pressing [V/MHz] will cycle through 1 MHz, 10 MHz or step size. Press [MENU] to exit the menu.

8. To set transmit tone, press [MENU]. Use [Up] / [Down] to navigate to DUP/TONE. Press [Enter]. Use [Up] / [Down] to select Repeater Tone. Press [Enter]. Use knob to adjust. Press [MENU] to exit the menu.

9. To set receive tone, press [MENU]. Use [Up] / [Down] to navigate to DUP/TONE. Press [Enter]. Use [Up] / [Down] to select TSQL Freq. Press [Enter]. Use knob to adjust. Press [MENU] to exit the menu.

10. To set subaudible tone type, press [QUICK]. Use [Up] / [Down] to navigate to TONE. Press [Enter]. Use [Up] / [Down] to select from TONE (CTCSS on transmit only), TSQL((.)) (CTCSS on receive only with beep), TSQL (CTCSS on receive only), DTCS((.)) (DCS with beep), DTCS (DCS), DTCS-R (reverse squelch DCS), DTCS(T) (DCS on transmit only), TONE(T)/DTCS(R) (CTCSS on transmit, DCS on receive), DTCS(T)-TSQL(R) (DCS on transmit, tone on receive) or OFF (no tones). Press [MENU] to exit the menu.

11. To set power level, hold [V/MHz] down and rotate the knob. This will cycle through SL0 (0.1 W), L01 (0.5 W), L02 (1 W), MID (2.5 W) and nothing displayed (5 W). Release [V/MHz] to set.

12. Hold [M/CALL] for one second to start write.

13. Scroll to desired memory using the knob.

14. Hold [M/CALL] for one second to write.

Lock/unlock radio

Hold [MENU] for one second to lock or unlock.

Check repeater input frequency

Hold [SQL] to listen on the repeater input.

Change power in the field

Hold [V/MHz] down and rotate the knob. This will cycle through SL0 (0.1 W), L01 (0.5 W), L02 (1 W), MID (2.5 W) and nothing displayed (5 W). Release [V/MHz] to set.

Adjust volume

Rotate ring to adjust volume.

Adjust squelch

Hold [SQL] and rotate knob to adjust squelch. Release [SQL] to set.

Weird Modes

Radio does not transmit

This radio has a PTT lock. Use [MENU], navigate to Function, select that and then select PTT Lock.

Useful Information

Hold [Power] for one second to turn off / on.

This radio has dual watch. Press [MAIN] to switch which side of the radio will transmit. Hold [MAIN] for one second to toggle dual watch on and off.

There are several version of this radio (ID-51, ID-51A, ID-51A Anniversary Edition, ID-51A Plus, ID-51A Plus2) with different colors. All are programmed similarly.

Factory reset

To reset, push [MENU]. Use [Down] to scroll to the last entry Other. Press [Enter] to select. Use [Up] / [Down] to select Reset, then press [Enter]. Use [Up] / [Down] to choose All Reset (clears everything). Press [Enter]. Press [Up] to select YES, then [Up] one more time to select YES again for a full reset.)

Settings reset

To reset, push [MENU]. Use [Down] to scroll to the last entry Other. Press [Enter] to select. Use [Up] / [Down] to select Reset, then press [Enter]. Use [Up] / [Down] to choose Partial Reset (clears settings but not memories). Press [Enter]. Press [Up] to select YES and the radio will reset.

Kenwood TH-78A

Radio Layout

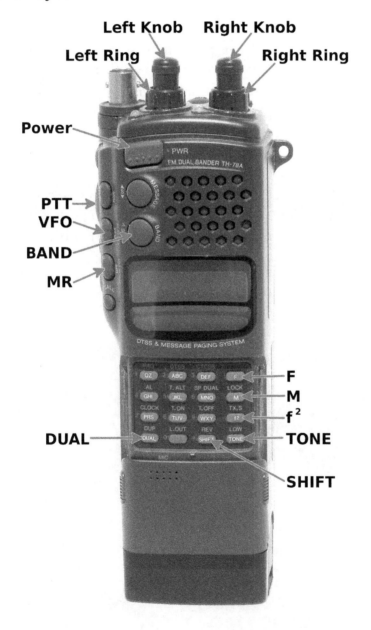

Specs

Receivers Two independent receivers, simultaneous receive
Receives 118–173.995 MHz FM, 438–449.985 MHz FM
Transmits 144–148 MHz FM @ 5 W, 438–450 MHz FM @ 5 W
Antenna connector SMA F on radio; needs SMA M antenna
Modes FM
Memory Channels 50, 150 with optional ME-1 memory expansion
Power 6.3–16 V DC, Egston 238 barrel style, 3.5mm OD, 1.3mm ID plug,
   center positive
Model year 1992

Standard Tasks

Program frequency in the field

1. Select a band by pressing [BAND].

2. Enter VFO mode by pressing [VFO].

3. Enter frequency using the number keypad, but leave off first digit (44390 is 144.390 MHz or 449.300 MHz depending on band).

4. To set repeater shift, press [SHIFT]. This will cycle through + (plus offset), – (minus offset) and no offset.

5. The radio has only global repeater offsets per band. To set repeater offset, first turn off the radio. Then hold down [SHIFT] while turning the radio on. (Display shows SHIFT.) Next, hold down [F] for at least one second, then release. (The display will blink F.) Next press [SHIFT]. Use the right knob to adjust, and press [SHIFT] to save. Note: If you see SPLIT instead of SHIFT, turn the radio off and turn it back on while holding down [SHIFT].

6. To enable CTCSS, press [TONE]. The display will show T (send tone on transmit). You can also set tone squelch by pressing [F] then [3/DEF].

7. To set CTCSS tone, hold down [F] for at least one second (F will blink). Press [TONE]. Then use the right knob to adjust. Press [TONE] to set.

8. To set power level, repeatedly press [F] then [TONE]. This will cycle through no power level display (high power, 5 W with PB-13 or PB-18 batteries, 2 W with PB-14, PB-17 or alkaline batteries), L (low power, 0.5 W) and EL (economy, 20 mW on 2m, 10 mW on 70cm).

9. To write memory, press [M]. Enter any two digit memory (01–49) then press [MR] to save.

10. Press [MR] to go to Memory mode. Use the right knob to select a memory.

Lock/unlock radio

Press and release $\boxed{F}$ and then $\boxed{M}$ to lock / unlock radio.

Check repeater input frequency

Press and release $\boxed{F}$ then $\boxed{SHIFT}$ to switch to repeater input frequency (R will be shown in display). Press $\boxed{F}$ then $\boxed{SHIFT}$ again to go back to normal operation.

Change power in the field

To set power level, repeatedly press $\boxed{F}$ then $\boxed{TONE}$. This will cycle through no power level display (high power, 5 W with PB-13 or PB-18 batteries, 2 W with PB-14, PB-17 or alkaline batteries), L (low power, 0.5 W) and EL (economy, 20 mW on 2m, 10 mW on 70cm).

Adjust volume

Rotate the left knob to adjust volume of the current side (0–20). (Press $\boxed{BAND}$ to switch sides.)

Adjust squelch

Rotate the left ring to adjust squelch of left (2m) band. Rotate the right ring to adjust squelch of the right (70cm) band.

Weird Modes

Can't set CTCSS Tone

Most radios sold in North America shipped with the TSU-7 option (tone board) installed. If your radio has no tone board, it will not transmit CTCSS tones.

Most buttons don't work

Assuming the radio is unlocked (no key icon) there is one other reason why keys might not work. This radio has a tone alert function that disables most of the keys. When on, you'll see a bell graphic in the display. Press $\boxed{F}$ and then button5/JKL to disable it.

No receive audio

This radio has a DTSS (dual-tone squelch) mode. When enabled, it waits for two DTMF tones before opening squelch. You can turn it off by pressing $\boxed{F}$ and then $\boxed{2/ABC}$.

No buttons

This radio has a sliding panel that protects the buttons from being pushed.
If the radio looks like this:

Panel closed

slide the panel down by pushing down on both the left and right sides at
the same time to expose the buttons.

Useful Information

If you have channel name turned on, you will be able to name channels
but will have half the number of memories you usually do. Hold down f^2
while turning the radio on to switch to/from this mode.

Press DUAL to switch between monitoring one frequency and monitoring both frequencies. Use BAND to determine which is the transmit
VFO.

To switch to full duplex operation, first switch to dual frequency display. Then press F for more than one second (F blinks) and then press
DUAL. The display will show a blinking DUP. Repeat to turn off.

This radio can operate as a crossband repeater. Hold down F for
more than one second (F blinks) and then press 0. Repeat the sequence
to disable.

Kenwood used a very hard-to-read red for the number buttons. The
keypad is:

1/QZ	2/ABC	3/DEF	F/A
4/GHI	5/JKL	6/MNO	M/B
7/PRS	8/TUV	9/WXY	f^2/C
#/DUAL	0	*/SHIFT	TONE/D

Factory reset

To reset everything including memories, hold down $\boxed{\text{M}}$ while turning on the radio.

Settings reset

To reset everything except memories, hold down $\boxed{\text{F}}$ while turning on the radio.

Kenwood TH-D7A

Radio Layout

Knob

Volume

PTT

Power

Moni

Esc

OK

A/B

Menu

F

VFO

MR

*/MHz   #/Ent

Specs

Receivers Two independent receivers, simultaneous receive
Receives 118–136 MHz AM (band A only), 136–173 MHz FM, 400–479
   MHz FM
Transmits 144–148 MHz FM @ 6 W, 430–450 MHz FM @ 5.5 W (using
   13.8V battery)
Antenna connector SMA F on radio; needs SMA M antenna
Modes FM
Memory Channels 200
Power 5.5–16 V DC, Egston 238 barrel style, 3.5mm OD, 1.3mm ID plug,
   center positive
Model year 1998 for A, 2000 for A(G)

Standard Tasks

Program frequency in the field

1. Select a VFO by pressing A/B.

2. Enter VFO mode by pressing VFO.

3. Press #/ENT to enter a frequency.

4. Enter frequency using the number keypad (144390 is 144.390 MHz).

5. To set repeater shift, press F then press */MHz. This will cycle
   through + (plus offset), – (minus offset) and no offset. The radio has
   automatic repeater shift and will usually guess correctly.

6. To set repeater offset, press F 5/LIST. Use the scroll knob to
   adjust. Press OK to save. Note that this offset will also be used for
   automatic repeater shift, so if you set a strange offset, put it back to
   normal after programming the memory.

7. To set CTCSS, press F 1/BAL to turn the tone on (radio shows T)
   or off.

8. To set CTCSS tone, press F 2/TNC. Use the scroll knob to adjust.
   Press OK to save.

9. To set power level, press F MENU. This will cycle through H (high,
   5 W), L (low, 0.5 W) and EL (economy, 50 mW).

10. To write memory, press F MR. Use the scroll knob to select a
    memory (empty memories have an empty triangle). Press OK to save.

11. Press MR to go to Memory mode. Use the scroll knob to select a
    memory.

Lock/unlock radio

Hold down F for two seconds to lock / unlock radio.

Check repeater input frequency

Press 7/REV to switch to repeater input frequency (R will be shown in display). Press 7/REV again to go back to normal operation.

Change power in the field

To set power level, press F MENU. This will cycle through H (high, 5 W), L (low, 0.5 W) and EL (economy, 50 mW).

Adjust volume

Rotate the outer volume ring to adjust volume.

Adjust squelch

Press F then press Moni. Use scroll knob to adjust squelch. Squelch is separate for each VFO.

Weird Modes

Radio doesn't transmit

This radio has a transmit inhibit function. To disable it, press MENU 1 5 5 then use the scroll knob to adjust to Off. Then press OK.

Radio shows Ch NN

This radio has a channel mode which shows memory numbers and prevents you from entering VFO mode. To disable or enable it, hold down A/B while turning radio on.

Can't hear one band

Volume for band A and band B can be adjusted so that one is louder than the other. Press 1/BAL then use scroll knob to change audio balance (higher is A louder, lower is B louder. At the extreme, the other band will be muted). Press OK to set.

Useful Information

This radio also comes in TH-D7A(G) and TH-D7E(G) models. These models have additional APRS functions.

Press 0/DUAL to switch between monitoring one frequency and monitoring both frequencies. Use A/B to determine which is the transmit VFO.

The ports on the right hand side (from top to bottom) are: earphone (2.5 mm stereo), microphone (3.5 mm stereo), PC cable (2.5 mm stereo), APRS communications (2.5 mm stereo) and 13.8 V DC input (Egston 238 barrel).

Factory reset

To reset the radio, hold down [F] while turning on the radio. Screen will go dark, then show AUX RESET?. Use scroll knob to choose FULL RESET (resets everything). Press [OK]. Use scroll knob to select YES to do the reset, then press [OK] to reset, or [ESC] to abort.

VFO reset

To reset the radio, hold down [F] while turning on the radio. Screen will go dark, then show AUX RESET?. Use scroll knob to choose VFO RESET (resets VFOs and their settings). Press [OK]. Use scroll knob to select YES to do the reset, then press [OK] to reset, or [ESC] to abort.

Kenwood TH-D72A

Radio Layout

Specs

Receivers Two independent receivers, simultaneous receive
Receives 136–174 MHz FM on VFO A, 410–470 MHz FM on VFO A, 118–
174 MHz FM on VFO B, 320–524 MHz FM on VFO B
Transmits 144–148 MHz FM @ 5 W, 430–450 MHz FM @ 5 W
Antenna connector SMA F on radio; needs SMA M antenna
Modes FM
Memory Channels 1000
Power 12–16 V DC, Egston 238 barrel style, 3.5mm OD, 1.3mm ID plug,
center positive
Model year 2010

Standard Tasks

Program frequency in the field

1. Select a VFO by pressing [A/B].

2. Enter VFO mode by pressing [VFO].

3. Press [ENT] to enter a frequency.

4. Enter frequency using the number keypad (144390 is 144.390 MHz).

5. To set repeater shift, press [F] then press [MHz]. This will cycle
   through + (plus offset), – (minus offset) and no offset. The radio has
   automatic repeater shift and will usually guess correctly.

6. To set repeater offset, press [MENU] [1] [6] [0]. Use the scroll knob
   or [Up] / [Down] to adjust. Press [OK] to save. Press [Menu] to exit the
   menu mode.

7. To set CTCSS or DCS mode, press [TONE]. The radio will cycle
   through T (send tone on transmit), CT (tone on transmit and recieve),
   DCS (digital coded squelch), D.0 (mixed DCS and CTCSS), and off.

8. To set CTCSS or DCS tone (whichever is enabled), press [F] [TONE].
   Use the scroll knob or [Up] / [Down] to adjust. Press [OK] to save.

9. To set power level, press [F] [MENU]. This will cycle through H (high,
   5 W), L (low, 0.5 W) and EL (economy, 50 mW).

10. To write memory, press [F] [MR]. Use the scroll knob or [Up] / [Down]
    to select a memory (empty memories have an empty triangle). Press
    [OK] to save.

11. Press [MR] to go to Memory mode. Use the scroll knob or [Up] / [Down]
    to select a memory.

Lock/unlock radio

Hold down [F] for two seconds to lock / unlock radio.

Check repeater input frequency

Press REV to switch to repeater input frequency (R will be shown in display). Press REV again to go back to normal operation.

Change power in the field

To set power level, press F MENU. This will cycle through H (high, 5 W), L (low, 0.5 W) and EL (economy, 50 mW).

Adjust volume

Rotate the outer volume ring to adjust volume.

Adjust squelch

Press F then press Moni. Use Up / Down or scroll knob to adjust squelch. Squelch is separate for each band.

Weird Modes

Can't set CTCSS Tone / DCS

This is disabled if the radio is not already set up in CTCSS or DCS mode. Use the TONE button to make sure you're in the right mode first.

Radio doesn't transmit

This radio has a transmit inhibit function. To disable it, press MENU 1 3 9 then use the scroll knob to adjust to Off. Then press OK.

Radio shows Ch NN

This radio has a channel mode which shows memory numbers and prevents you from entering VFO mode. To disable or enable it, hold down PTT and A/B while turning radio on. (Does not transmit.) Resetting the radio is not possible in channel mode.

Useful Information

Press DUAL to switch between monitoring one frequency and monitoring both frequencies. Use A/B to determine which is the transmit VFO.

The ports on the right hand side (from top to bottom) are: earphone (2.5 mm stereo), microphone (3.5 mm stereo), mini-USB for programming, APRS communications (2.5 mm stereo) and 13.8 V DC input (Egston 238 barrel).

Volume for VFO A and VFO B can be adjusted so that one is louder than the other. Press MENU 1 2 0 then use Up / Down or scroll knob to change audio balance. Press OK to set.

Factory reset

To reset the radio, hold down [F] while turning on the radio. Use [Up] / [Down] or scroll knob to choose `Full Reset` (resets everything). Press [OK] to reset, or [ESC] to abort.

Settings reset

To reset the radio, hold down [F] while turning on the radio. Use [Up] / [Down] or scroll knob to choose `Partial Reset` (resets almost everything except memory channels). Press [OK] to reset, or [ESC] to abort.

VFO reset

To reset the radio, hold down [F] while turning on the radio. Use [Up] / [Down] or scroll knob to choose `VFO Reset` (resets VFOs and their settings). Press [OK] to reset, or [ESC] to abort.

Kenwood TH-D74A

Radio Layout

Specs

Receivers Two independent receivers, simultaneous receive
Receives 136–174 MHz FM / AM / SSB /DSTAR, 216–220 MHz FM / AM / SSB / DSTAR, 410–470 MHz FM / AM / SSB / DSTAR all on band A, 0.1–524 MHz on band B
Transmits 144–148 MHz FM / DSTAR @ 5 W, 222–225 MHz FM / DSTAR @ 5 W, 430–450 MHz FM / DSTAR @ 5 W; transmit on band A only
Antenna connector SMA F on radio; needs SMA M antenna
Modes FM, DSTAR
Memory Channels 1000
Power 11.0–15.9 V DC, Egston 238 barrel style, 3.5mm OD, 1.3mm ID plug, center positive
Model year 2016

Standard Tasks

Program frequency in the field

1. Press A/B until you see one frequency only.

2. Press Mode until you see FM.

3. Press Ent.

4. Enter frequency (144390 for 144.390 MHz).

5. To set up tone type, press 8 (TONE) to cycle through T (send tone on transmit), CT (cross-tone, different up and down), DCS (digital coded squelch), D/O (DCS up, carrier squelch down) and blank (off). You might also see T/C (tone up, CTCSS down), D/C (DCS up, CTCSS down) or T/D (tone up, DCS down).

6. To select tone, press F 8 (Tone). Use the knob to adjust the value. Press A/B to set.

7. To change repeater shift, press F 7 (Shift) to cycle through + (positive shift), – (negative shift) and blank (no shift).

8. If you need to change offset frequency, press Menu 1 4 0. Use the knob to adjust the offset frequency, and press A/B to set. Press Menu to exit.

9. To set power level, press F Menu. This will cycle through H (high, 5 W), M (medium, 2 W), L (low, 0.5 W) and EL (economy, 50 mW).

10. To write the memory, press F 2 (MR) then use the knob or direct keypad entry to select the memory desired. Press Ent to write. Press Left to go back to the main screen.

11. Press 2 (MR) to go to Memory mode. Use the knob to select a memory.

Lock/unlock radio

Hold down [F] for two seconds to lock / unlock radio.

Check repeater input frequency

Press [7] (Rev) to switch to repeater input frequency (R will be shown in display). Press [7] (Rev) again to go back to normal operation.

Change power in the field

To set power level, press [F] [Menu]. This will cycle through H (high, 5 W), M (medium, 2 W), L (low, 0.5 W) and EL (economy, 50 mW).

Adjust volume

Rotate the outer volume ring to adjust volume.

Adjust squelch

Press [F] [Moni] and use knob to adjust: Open or 1–5. Press [Ent] to set.

Weird Modes

Can't set CTCSS Tone / DCS

This is disabled if the radio is not already set up in CTCSS or DCS mode. Use the [2] (Tone) button to make sure you're in the right mode first.

Radio doesn't transmit

This radio has a transmit inhibit function. To disable it, press [Menu] [1] [1] [0] then use the knob to adjust to Off. Then press [Ent].

Can't hear one side of the radio

Volume for VFO A and VFO B can be adjusted so that one is louder than the other, or so that one is entirely silent. Press [Menu] [9] [1] [0] then use the knob to change audio balance. Press [Ent] to set.

Radio display is black or has blinking H

The radio is getting hot or has overheated. Wait for it to cool down before transmitting again.

Useful Information

Press [Power] for one second to turn the radio on or off.

Press [F] [A/B] to switch between monitoring one frequency and monitoring both frequencies.

This radio has a MicroSD card slot on the right hand side under the speaker mic connector. Under that is a micro USB connector.

Factory reset

To reset the radio, hold down [F] while turning on the radio. Use knob to choose `Full Reset` (resets everything). Press [A/B] to reset.

Settings reset

To reset the radio, hold down [F] while turning on the radio. Use knob to choose `Partial Reset` (resets almost everything except memory channels). Press [A/B] to reset.

VFO reset

To reset the radio, hold down [F] while turning on the radio. Use knob to choose `VFO Reset` (resets VFOs and their settings). Press [A/B] to reset.

Kenwood TH-F6A

Radio Layout

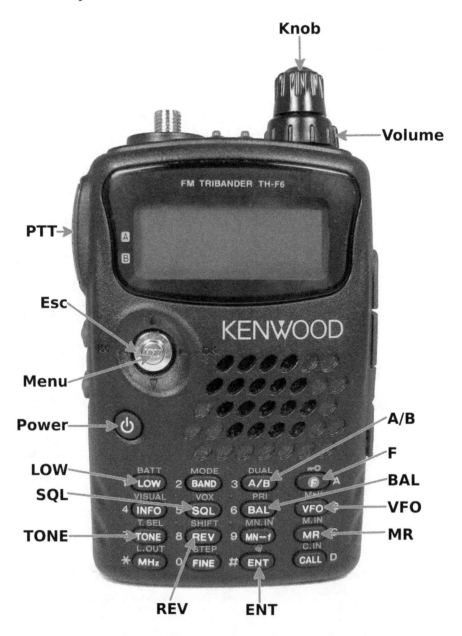

Specs

Receivers Two independent receivers, simultaneous receive
Receives 136–174 MHz FM on VFO A, 216–260 MHz FM on VFO A, 410–470 MHz FM on VFO A, 0.1–470 MHz AM/FM/SSB on VFO B, 470–824 MHz FM on VFO B, 849–869 MHz FM on VFO B, 894–1300 MHz on VFO B
Transmits 144–148 MHz FM @ 5 W, 222–225 MHz FM @ 5 W, 430–450 MHz FM @ 5 W
Antenna connector SMA F on radio; needs SMA M antenna
Modes AM, FM, SSB
Memory Channels 400
Power 12–16 V DC, Egston 238 barrel style, 3.5mm OD, 1.3mm ID plug, center positive
Model year 2001

Standard Tasks

Program frequency in the field

1. Select a VFO by pressing A/B.

2. Enter VFO mode by pressing VFO.

3. Press ENT to enter a frequency.

4. Enter frequency using the number keypad (144390 is 144.390 MHz).

5. To set repeater shift, repeatedly press F then REV. This will cycle through + (plus offset), – (minus offset) and no offset. The radio has automatic repeater shift and will usually guess correctly.

6. To set repeater offset, press Menu then use the knob to scroll to menu 6, OFFSET. Press Menu then use the knob to change value. Press Menu to store. Press Esc to exit menu mode.

7. To set CTCSS or DCS mode, press TONE. The radio will cycle through T (send tone on transmit), CT (tone on transmit and recieve), DCS (digital coded squelch), and off.

8. To set CTCSS or DCS tone (whichever is enabled), press F TONE. Use the knob to adjust. Press Menu to save, then Esc to exit menu mode.

9. To set power level, press LOW. This will cycle through L (low, 0.5 W), EL (economy, 50 mW) and H (high, 5 W).

10. To write memory, press F. Use the knob to select a memory (empty memories have an empty triangle). Press Menu to save.

11. Press MR to go to Memory mode. Use the knob to select a memory.

Lock/unlock radio

Hold down ⌊F⌋ for two seconds to lock / unlock radio.

Check repeater input frequency

Press ⌊REV⌋ to switch to repeater input frequency (R will be shown in display). Press ⌊REV⌋ again to go back to normal operation.

Change power in the field

To set power level, press ⌊LOW⌋. This will cycle through L (low, 0.5 W), EL (economy, 50 mW) and H (high, 5 W).

Adjust volume

Rotate the outer volume ring to adjust volume.

Adjust squelch

Press ⌊SQL⌋ then use knob to adjust squelch. Squelch is separate for each band.

Weird Modes

Can't set CTCSS Tone / DCS

This is disabled if the radio is not already set up in CTCSS or DCS mode. Use the ⌊TONE⌋ button to make sure you're in the right mode first.

Radio doesn't transmit

This radio has a transmit inhibit function. To disable it, press ⌊Menu⌋ then scroll to menu 8, TX INHIBIT. Press ⌊Menu⌋ then use the knob to adjust to OFF. Press ⌊Menu⌋ to save, then press ⌊Esc⌋ to exit.

Radio displays 00:00 or some other time

This radio has "Tone Alert," which displays the time since a signal was received. When it is enabled, many things don't work, and it will beep madly if someone transmits. Press ⌊F⌋ then ⌊ENT⌋ to disable it.

Speaker mic doesn't work

This radio needs to be told what the speaker mic jacks are being used for. Press ⌊Menu⌋ and use the knob to go to menu 9, SP/MIC JACK Press ⌊Menu⌋ and use the knob to select from SP/MIC (speaker mic), TNC (external TNC) and PC (PC programming mode). Press ⌊Menu⌋ to select, then ⌊Esc⌋ to exit.

Radio shows Ch NN

This radio has a channel mode which shows memory numbers and prevents you from entering VFO mode. To disable or enable it, hold down [A/B] while turning the radio on.

Don't hear one receiver

Volume for VFO A and VFO B can be adjusted so that one is louder than the other. Press [BAL] then use knob to change audio balance. Press [Menu] to set.

Useful Information

Press [F] then [A/B] to switch between monitoring one frequency and monitoring both frequencies. Use [A/B] to determine which is the transmit VFO.

The [Menu] button is a multi-way button. Pushing straight down is [Menu]. Pushing left is [Esc]. Pushing up and down can be used instead of the knob in many cases. Pushing right is sometimes called "OK."

The TH-F6A and the TH-F7E are roughly identical radios (even sharing the same manual); the TH-F7E was made for Europe and lacks the 1.25m band.

Factory reset

To reset the radio, hold down [F] while turning on the radio. Use the knob to choose FULL RESET (resets everything). Press [Menu] to select, or [Esc] to abort.

Settings reset

To reset the radio, hold down [F] while turning on the radio. Use the knob to choose MENU RESET (resets almost everything except memory channels). Press [Menu] to select, or [Esc] to abort.

VFO reset

To reset the radio, hold down [F] while turning on the radio. Use the knob to choose VFO RESET (resets VFOs and their settings). Press [Menu] to select, or [Esc] to abort.

Kenwood TH-K20A

Radio Layout

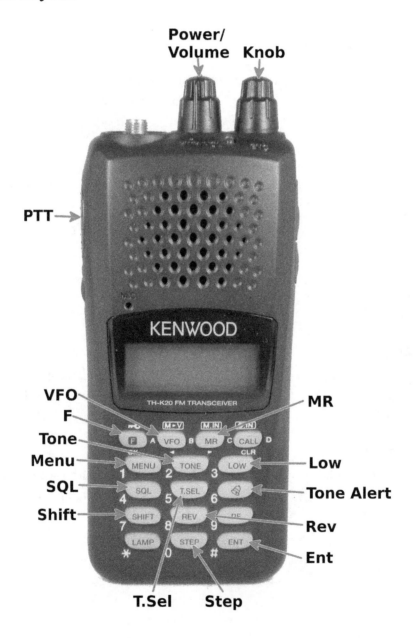

Specs

Receivers Single receiver with priority channel option
Receives 136–174 MHz FM
Transmits 144–148 MHz FM @ 5.5 W
Antenna connector SMA F on radio; needs SMA M antenna
Modes FM
Memory Channels 200
Power No DC input on radio, 7.4–9 V DC based on battery options
Model year 2011

Standard Tasks

Program frequency in the field

1. Enter VFO mode by pressing [VFO].

2. Enter frequency by pressing [Ent] then using the number keypad (144390 is 144.390 MHz). If needed you can adjust step size with [Step] (0).

3. To set repeater shift, press [Shift] (7). This will cycle through + (plus offset), – (minus offset) and no offset.

4. If you need to set the repeater offset frequency, press [Menu] and use the right knob to select menu 7, OFFSET. Press [F]. Use the knob to select the offset. Press [F] to set, and then press [Menu] to exit.

5. To set the tone mode, press [Tone] to select from T (CTCSS tone on transmit), CT (CTCSS tone on transmit and receive), DCS (digital coded squelch) and a triangle symbol (cross tone: separate tones for transmit and receive).

6. To set the tone value, press [T.Sel] (5). Then use the right knob to adjust tone value, and press [T.Sel] (5) to set.

7. To set power level, repeatedly press [Low] (3). This will cycle through no power level display (high power, 5.5 W), M (medium power, 2 W), and L (low power, 1 W).

8. To write memory, press [F] [MR]. Use the right knob to scroll to the memory to write, then press [MR] again to write.

9. Press [MR] to go to Memory mode. Use the right knob to select a memory.

Lock/unlock radio

Hold [F] for one second to lock / unlock radio.

Check repeater input frequency

Press [Rev] (8). Display will show R. Press [Rev] again to return to normal operation.

Change power in the field

To set power level, repeatedly press [Low] (3). This will cycle through no power level display (high power, 5.5 W), M (medium power, 2 W), and L (low power, 1 W).

Adjust volume

Rotate the power/volume (left) knob to adjust volume.

Adjust squelch

Press [SQL] (4) to adjust squelch. Rotate the right knob to select from 0–5. Press [SQL] to set.

Weird Modes

Can't enter VFO mode

This radio has a channel-only mode. Hold down [PTT] (does not transmit) and [MR] while turning the radio on to exit this mode.

Most buttons don't work

Assuming the radio is unlocked (no key icon) and not in channel mode, there is one other reason why keys might not work. This radio has a tone alert function that disables most of the keys. When on, you'll see a bell icon in the display. To disable, press the [Tone Alert] button, use the knob to scroll to OFF and press [Tone Alert] to set.

Radio doesn't transmit

The Kenwood TH-K20A has a transmit inhibit function. To disable it, press [Menu] and use the right knob to select menu 21, TX.INH. Press [F]. Use the knob to select OFF. Press [F] to set, and then press [Menu] to exit.

Useful Information

The Kenwood TH-K20A is also available in a 70cm version, the Kenwood TH-K40A. Programming is similar.

Factory reset

hold down ⌐F⌐ while turning the radio on. Use the knob to select FL.RST.? (full). Press ⌐F⌐. You will be prompted with SURE ?. Press ⌐F⌐ to reset. The radio must be unlocked to reset.

Settings reset

hold down ⌐F⌐ while turning the radio on. Use the knob to select PA.RST.? (partial). Press ⌐F⌐. You will be prompted with SURE ?. Press ⌐F⌐ to reset. The radio must be unlocked to reset.

Pofung

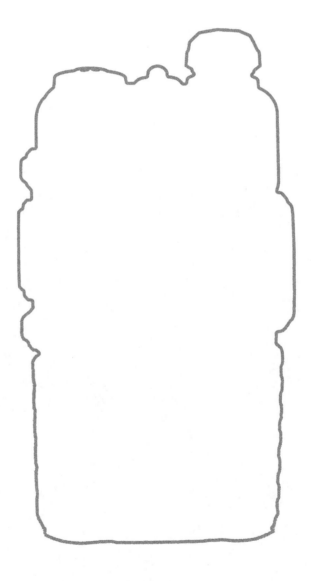

See Baofeng

Pofung: see Baofeng

On May 7, 2014, Baofeng re-branded its products as Pofung for international markets. This was to make the name easier to pronounce in non-Chinese markets. The two brands cover the same product.

Because Chinese versions of the radios are still available around the world, it is likely that both brands will be widespread.

Racal

See Thales

Racal: see Thales

On January 13, 2000, Racal was purchased by Thomson-CSF. The merged companies were renamed Thomson-CSF Racal plc. On December 6, 2000, the company was again renamed to Thales Group.

Radio Shack HTX-200

Radio Layout

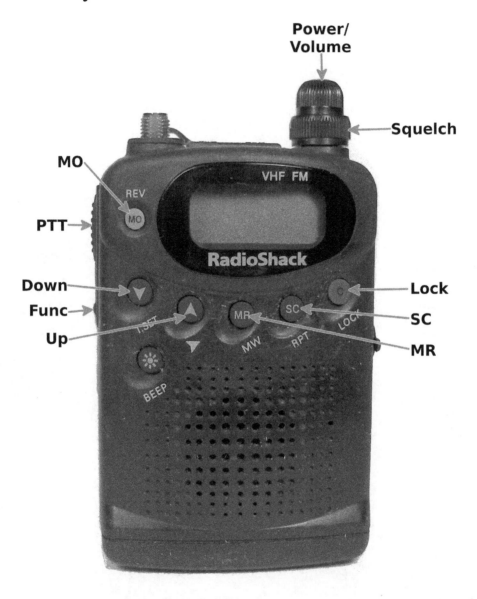

Specs

Receivers Single receiver
Receives 136–174 MHz FM
Transmits 142–149.885 MHz @ 200 mW FM
Antenna connector SMA F on radio; needs SMA M antenna
Modes FM
Memory Channels 30
Power 9 V DC, "I" barrel style jack 3.8 mm OD, 1.1 mm ID center positive
Model year 1998

Standard Tasks

Program frequency in the field

1. Press [MR] to go to memory mode if you aren't there already. (Screen will show MR in the upper right.) Use [Up] and [Down] to select the memory you want to write to.

2. Hold down [Func] and press [MR]. The memory will flash.

3. To enter a frequency, hold down [Func] and press [Up]. Press [Up] or [Down] to change digit. Hold down [Func] and press [Up] to move to next digit.

4. Hold down [Func] and press [SC] to set repeater offset in MHz (rPt). Use [Up] and [Down] to select it. (600 kHz is 0.6.) Use offset of 0.0 for simplex.

5. Hold down [Func] and press [MO] to set repeater shift (+,-) which will be displayed above the second digit in the frequency.

6. To turn CTCSS tone or tone squelch on, hold down [Func] and press [Down]. Display will show tONE oF. Press [Down] until display shows tONE oN. Next hold down [Func] and press [Down]; you will see rC (receive tone). Adjust it to your value (or oFF) using [Up] and [Down]. Then hold down [Func] and press [Down] to see tC (transmit tone). Adjust it to your value using [Up] and [Down].

7. Press [PTT] to write the memory. Note that if you leave memory mode (stops flashing) you will have to re-enter it by holding down [Func] and pressing [MR]. If you hold [Func] down for too long without pressing another key, the memory will be deleted.

Lock/unlock radio

Hold down [Func] and press [Lock] to lock/unlock. This locks everything except [PTT], volume, squelch and backlight. Key symbol will be displayed when radio is locked.

Check repeater input frequency

Hold down Func and press PTT (does not transmit). Display will change to the input frequency. There is no other indication that you're in reverse, so be cautious. Repeat procedure to switch back.

Change power in the field

There is no setting for power level. Power is 200 mW on batteries, or 2 W on 9V DC.

Adjust volume

Rotate inner power/volume knob to adjust volume.

Adjust squelch

Rotate squelch knob (outer knob).

Weird Modes

Display shows all LCD segments or PrESS

This radio has a diagnostic mode that you can enter by pushing MO while turning on the radio. If the display is showing all LCD segments or PrESS followed by a few letters, turn the radio off and then on again to skip the diagnostic.

Display shows EEP-Error

The EEPROM info is bad. You might be able to reset the radio using the reset procedure described below.

Display shows PLL-Error

The PLL couldn't get a lock. Power off and power on again.

Display shows S-SHORT

The external microphone has a short. Remove the external mic and use the internal mic.

Display shows Inhibit on PTT

You are trying to transmit outside of the range allowed by the radio. Check your offset.

Radio doesn't transmit

The radio has a busy channel lock-out mode which prevents transmission when there is a carrier on the frequency. If it has been enabled, you can disable it with the following procedure. Turn the power on while holding

Func. Hold down Func and press Down until you see bCLO-oN. Press Down to turn it off. Press PTT to set (does not transmit).

Useful Information

Holding down MO will open squelch for as long as you hold it down.

You can see the current receive tone (rc), transmit tone (tc), memory lockout (SCSP), repeater offset MHz (rpt) and frequency step MHz (CS) by holding the MO button down for more than one second.

By default, the radio transmits from 144.0 through 148.0 MHz. To expand the transmit range to 142.0 through 149.885, hold down SC when turning on the radio. The range will be shown briefly in the display. Repeating the operation restricts the transmit range again.

The configuration menu you get when you hold down Func and turn on the power has the following options: CS (frequency step), bCLO (busy channel lock-out), t_dY (drop CTCSS tone before ending transmission to prevent squelch tail), Sd (scan delay in seconds), tot (time-out timer in seconds), PS (power save), rPt (VFO repeater shift). Hold down Func and press Down to cycle through them. Press PTT to exit (does not transmit).

Programming the VFO for repeater operation is a little tricky (need to set VFO offset in config menu) and not necessary. Just write a memory, and use the VFO for simplex only.

Factory reset

To reset the radio, hold down Func and MO while turning the radio on. The display will show InitiAL as it resets. This also clears all memories.

Radio Shack HTX-202

Radio Layout

(Image is of 1994 model)

Specs

Receivers Single receiver
Receives 144–148 MHz FM
Transmits 144–148 MHz @ 6 W FM
Antenna connector BNC F on radio; needs BNC M antenna
Modes FM
Memory Channels 12 standard
Power 7.2–13.8 V DC per spec, "K" barrel style jack 5 mm OD, 2.1 mm ID
center positive
Model year 1992 (Realistic markings), 1994 (Radio Shack markings)

Standard Tasks

Program frequency in the field

1. Press ⎡D⎤ (VF) to go to VFO mode if you aren't there already. (Screen will show M-CH if you aren't.)

2. Enter the last four digits of the frequency on the numeric keypad (so 144.390 is 4390).

3. For repeater shift, hold down ⎡F⎤ and press ⎡3⎤ (+/-) until correct direction appears. (Note that this radio gets repeater shift wrong for 147 MHz, so be careful.)

4. To turn CTCSS tone or tone squelch on, hold down ⎡F⎤ and press ⎡1⎤ (T-SQL). Display will show T-SQL.

5. Offset defaults to 600 kHz. If you need a different offset, hold down ⎡F⎤ and press ⎡8⎤ (M-SET). Display will show oS. Use the right knob to adjust tone; default is 0.600. Press ⎡PTT⎤ to set (does not transmit).

6. Now write the frequency to a memory. Hold down ⎡F⎤ and turn the right knob until the memory number you want to write to is displayed. Release ⎡F⎤. Hold down ⎡F⎤ and then press ⎡C⎤ (M-WR) for one second to write.

7. Press ⎡C⎤ (MR) to go to memory mode.

8. To set repeater tone, hold down ⎡F⎤ and press ⎡8⎤ (M-SET). Radio will show tF (transmit frequency). Press ⎡*⎤ (down SC) to see tc (transmit tone). Rotate the right knob to set the value or use oFF for no transmit tone. Press ⎡*⎤ (down SC) to see rc (receive squelch tone). Rotate the right knob to set the value or use oFF for no receive squelch tone. Press ⎡PTT⎤ to set (does not transmit).

9. For power level, press the Low Power switch. Switch down is low power (about 1 W); switch up is high power (between 2.5 W and 6 W depending on power source).

Lock/unlock radio

Hold down ⎡F⎤ and press ⎡A⎤ (LOCK) to lock or unlock. The lock disables everything except PTT, Volume and Squelch.

Check repeater input frequency

Hold down ⎡F⎤ and press ⎡6⎤ (REV). Display will switch from receive frequency to transmit frequency, and repeater shift will change from + to - or vice versa. Repeat procedure to switch back.

Change power in the field

Press the Low Power switch. Switch down is low power (about 1 W); switch up is high power (between 2.5 W and 6 W depending on power source).

Adjust volume

Rotate the center power/volume knob to adjust volume.

Adjust squelch

Rotate the left Squelch knob.

Weird Modes

Display shows ER1

This means the lithium battery that maintains memory when the radio is powered off has died. Use the reset procedure described below to bypass the error condition.

Display shows ER2

This means the PLL is unable to lock. The radio will need repair.

Radio doesn't transmit

The HTX-202 can be set to prevent transmission. If it's set, the lower left corner of the LCD will display INH. To turn transmissions back on, make sure you're in VFO mode by pressing ⎡D⎤ (VF). Then hold down ⎡F⎤ and press ⎡8⎤ (M-SET). Press ⎡*⎤ (down SC) eleven times until you see tE. Use the right knob to adjust between oFF (no transmission) and on (transmission allowed). Press ⎡PTT⎤ to set (does not transmit).

Radio doesn't receive

The HTX-202 can be set to require a DTMF sequence before it opens the squelch. If this is turned on, DTMF will appear in the lower right of the display. To turn it off, hold ⎡F⎤ and press ⎡4⎤ (D-SQL).

Useful Information

You can see the transmit frequency, tone and tone squelch values by holding the M button (above the PTT and below F).

In VFO mode, the menu you get when you hold down F and press 8 (M-SET) has the following options: oS (offset), tc (default transmit tone value), rc (default receive tone squelch value), Sr (frequency step), Sc (scan resume), Sd (scan delay), S1 (lower scan limit), S2 (upper scan limit), ud (scan direction), PS (power save—higher number is less power used), tE (transmit enable), to (transmit timeout), Lb (priority channel scan time), Ar (touch tone auto reply). In memory mode, you'll see tF (transmit frequency for current memory), tc (transmit tone value for current memory) and rc (receive tone squelch for current frequency).

⚠ **WARNING:** The belt clip of the HTX-202 also doubles as the heat sink for the final transistors. Don't operate without the belt clip in place.

⚠ **WARNING:** Always remove the battery pack before operating on external power.

The 1992 model with Realistic markings resets to 146.200; the 1994 model with Radio Shack markings resets to 146.600.

Factory reset

To reset the radio (and clear an ER1 condition), hold down D (CLR) and F while turning the radio on. This also clears all memories and settings.

Radio Shack HTX-245

Radio Layout

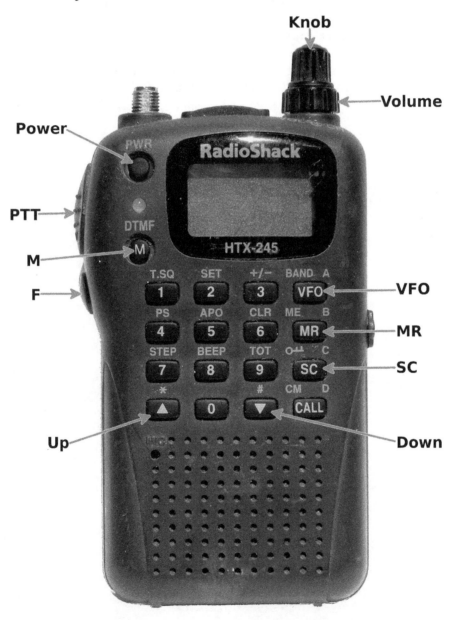

Specs

Receivers Single receiver
Receives 144–148 MHz FM, 438–450 MHz FM
Transmits 144–148 MHz @ 1.5 W FM, 438–450 MHz @ 1.5 W FM
Antenna connector SMA F on radio; needs SMA M antenna
Modes FM
Memory Channels 50
Power 4.5–6 V DC, "H" barrel style jack 3.4 mm OD, 1.3 mm ID center
   positive
Model year 2000

Standard Tasks

Program frequency in the field

1. Press [VFO] to go to VFO mode if you aren't there already. (Screen will show MR if you aren't.)

2. Select band by holding [F] and pressing [VFO] to scroll through 2 m, NOAA weather and 70 cm.

3. Enter frequency on numeric keypad (144.390 is entered as 144390).

4. For repeater shift, hold down [F] and press [3] (+/-) until correct direction appears (+ is positive, - is negative, blank is simplex).

5. To turn CTCSS tone or tone squelch on, hold down [F] and press [1] (T-SQL) to cycle through T (tone), TSQ (tone squelch) and blank (no tone).

6. To set repeater tone, hold down [F] and press [2] (SET). Display will show Sq (looks like 59 to most of us). Use knob to scroll to rt (receive tone) and/or tt (transmit tone). Use [Up] and [Down] to adjust.

7. There are two offsets in the radio. One defaults to 600 kHz; the other to 5 MHz. If you need a different offset, while in set mode use the knob to scroll to r1 (2 m offset, default 0.60) or r2 (70 cm offset, default 5.00). Use [Up] and [Down] to adjust.

8. Hold [F] and press [2] to exit.

9. Now write the frequency to a memory. Hold down [F] then press and release [MR]. Turn knob until the memory number you want to write is displayed. Hold down [F] and press [MR] to write.

10. Press [MR] to go to memory mode. You should be on the channel you just wrote; use knob to change if you want.

Lock/unlock radio

Hold down [F] and press [SC] to lock or unlock. The lock disables everything except PTT, volume, monitor and power.

Check repeater input frequency

There is no way to listen to the repeater input frequency except to program a separate memory for it.

Change power in the field

This radio has only one power level. On 6V DC external supply it provides 1.5 W output. On 4.5V DC internal batteries it provides 700 mW.

Adjust volume

Rotate the ring to adjust volume.

Adjust squelch

Press [VFO], then hold [F] and press [2] to enter Set mode. Rotate knob until you see $q (looks like 59 to most of us). Press [Up] and [Down] to set it from 0 (no squelch) to 5 (highest squelch). Hold [F] and press [2] to exit.

Weird Modes

Can't change parameters

Some parameters require you to be in VFO mode to change them. Press [VFO] before pressing the buttons you want.

Useful Information

As well as monitoring, the [M] button will display the transmit frequency, tone and tone squelch values when you hold it down.

⚠ **WARNING:** There is a switch in the battery compartment which controls whether or not an external adapter charges the batteries. For NiCd or NiMH, the power switch should be set to "On." For other batteries, the power switch should be set to "Off."

This radio has two versions, part numbers 19-1106 and 19-1106A. The "A" version can be given extended transmit/receive frequency ranges (142.00–149.88 MHz and 420–450 MHz). To do this, hold down [2] and [3] and turn the radio on. The non-"A" version requires a hardware mod.

Factory reset

To reset the radio hold down [6] (CLR) and [MR] while turning the radio on. This clears all memories and settings.

Radio Shack HTX-420

Radio Layout

Specs

Receivers Single receiver with priority channel option
Receives 108–136.9875 MHz AM, 137–174 MHz FM, 420–512 MHz FM
Transmits 144–148 MHz @ 4 W FM, 438–450 MHz @ 4 W FM
Antenna connector SMA F on radio; needs SMA M antenna
Modes FM
Memory Channels 100
Power 5.0–13.8 V DC, "H" barrel style jack 3.4 mm OD, 1.3 mm ID center
  positive
Model year 2002

Standard Tasks

Program frequency in the field

1. Press MR to go to VFO mode if you aren't there already. (Screen will show MR if you aren't.)

2. Select band by pressing Middle repeatedly to go through 2 m, 70 cm, air band and NOAA weather.

3. Hold F and press 2 (STEP) to adjust step if necessary. Use Up and Down to select step, and Middle to set.

4. Enter frequency on numeric keypad (144.390 is entered as 144390).

5. For repeater shift, hold down F and press 9 (+/-) until correct direction appears (+ is positive, - is negative, blank is simplex).

6. To set CTCSS tone for transmit or receive, press Right until you see RXTONE or TXTONE. Then press Up or Down to adjust, then Middle to set.

7. To set repeater tone, hold down F and press 1 (T.SQL). This will cycle through T (transmit tone), SQ (receive tone), TSQ (both send and receive tone) and blank (no tone).

8. There are two offsets in the radio, one for VHF frequencies and another for UHF frequencies. To change these values, press Right until you see VHFRPT or UHFRPT. Then press Up or Down to adjust, and Enter to set.

9. To set power level, press H/L. This will cycle through L (1 W), M (approximately 1.5 W on batteries, 2 W on external power) and blank (high power: 3 W on batteries, 4 W on external power).

10. Now write the frequency to a memory. Hold down F then press and release MR. Use Up or Down or the knob until the memory number you want to write is displayed. Hold down F and press MR to write.

11. Press MR to go to memory mode. You should be on the channel you just wrote; use knob to change if you want.

Lock/unlock radio

Hold down F and press DW to lock or unlock. The lock disables everything except PTT, volume, monitor and power.

Check repeater input frequency

Hold down F and press 8 (REV) to switch to reverse mode (display shows opposite offset, but otherwise gives no indication you're in reverse). Hold F and press 8 to return to normal.

Change power in the field

To set power level, press H/L. This will cycle through L (1 W), M (approximately 1.5 W on batteries, 2 W on external power) and blank (high power, 3 W on batteries, 4 W on external power).

Adjust volume

Rotate the ring to adjust volume.

Adjust squelch

Press Right. Display will show SQL. Press Up and Down to adjust, then press Middle to set.

Weird Modes

Display shows ERR

The radio will show ERR when it loses PLL lock. This can happen if RF gets into the PLL circuit. Turn off the radio and turn it back on to clear it; if it happens regularly reduce power.

If the radio shows ERR at other times, it may need service.

Useful Information

The external power socket on the radio just powers the radio. The battery has its own power socket (same style) that charges the battery. The charging voltage is 12.0 V DC.

The radio can transmit on one frequency and monitor another for satellite operation. Press MR to go to VFO mode, then enter your receive frequency. Then Hold F and press SC (display shows XB). Next, enter your transmit frequency. You can hold F and press 8 to switch frequencies. Hold F and press SC to leave this mode.

This radio comes with a compass. To use it, hold down [F] and press [H/L] (COMP). The radio must be roughly horizontal to use the compass.

To calibrate the compass, place the radio flat on its back. Then hold down [H/L] while turning the power on. (Radio will show CAL.) Next, rotate the radio around twice, taking about ten seconds per rotation. Finally, push [Middle].

This radio has an extended transmit frequency range (142.000–149.880 MHz and 420.000-470.000 MHz). To do this, hold down [SC] and [9] while turning on the power. This will reset the radio (losing all memories/settings) as well as changing the frequency limits. Repeat the procedure to select normal frequency range.

Adjusting the volume on this radio can affect the knob and vice versa. Use care when moving one that you don't move the other.

Factory reset

To reset the radio hold down [F] and [6] (CLR) while turning the radio on. This clears all memories and settings.

Radio Layout

Power/Volume **Knob**

Circle **Diamond**

Square **Enter**

PTT

(Racal label)

Specs

Receivers Single receiver
Receives 136–174 MHz FM
Transmits 136–174 MHz @ 5 W FM, 2 W FM using AA batteries
Antenna connector SMA **M** on radio; needs SMA **F** antenna. Note that
    this radio requires an SF (Motorola-style) SMA connector
Modes FM, P25
Memory Channels 256, 304 on radios with Fire Feature
Power No DC input on radio; 8–14 V DC per service manual
Model year 1998

Standard Tasks

Program frequency in the field

1. Select the memory you want to program using the selector knob. (Depending on how the radio is programmed, you may be able to use the switch to select zones. You may also be able to go to different banks. See "Useful Info" below for details.)

2. Press Enter.

3. Press Diamond to scroll up to PROGRAM. Press Enter.

4. Press Circle to scroll down to CHANEL. Press Enter.

5. To set mode, press Circle to scroll down to MODE=. If that is not already ANALOG, press Enter to select, press Circle to scroll to ANALOG, then press Enter to set. (Your other choice is DIGITAL, which is P25.)

6. To set bandwidth, press Circle to scroll down to B/W. If that is not already 25 kHz, press Circle to scroll to 25 kHz, then press Enter to set. (Your other choice is 12.5 kHz.)

7. To set receive frequency, press Circle to scroll down to RX=. Enter the receive frequency using the number pad. (144390 for 144.390 MHz). Press Enter to set.

8. To set receive squelch, press Circle to scroll down to RXSQMD. Press Enter to change. Press Circle to scroll through squelch types. The options are NONE (no squelch), NOISE (carrier squelch), TONE (CTCSS squelch) and DCS (DCS squelch). Press Enter to set.

9. Press Circle to set squelch value, which will be SQ=, TONE= or CODE=. Scroll through the values using Circle and Diamond. Press Enter to set. (There are seventeen levels of squelch; the default is 3.)

10. To set transmit frequency, press Circle to scroll down to TX=. Enter the receive frequency using the number pad. (144390 for 144.390 MHz). Press Enter to set.

11. Press ⬚Circle⬚ to scroll down to TXSQMD=. Press ⬚Enter⬚ to change; your options are NONE, CTCSS and DCS. Press ⬚Enter⬚ again to set.

12. Press ⬚Circle⬚ to scroll down to TON= (or CODE= for DCS). Press ⬚Enter⬚ to change, then scroll with ⬚Circle⬚ and ⬚Diamond⬚. Press ⬚Enter⬚ to set.

13. For low power level, press ⬚Circle⬚ to scroll down to LO PWR= and press ⬚Enter⬚ to set. Use Circle and Diamond to adjust. You can choose from 0.1 W, 0.5 W, 1.0 W, 2.0 W and 5.0 W. Press ⬚Enter⬚ to set.

14. For high power level, press ⬚Circle⬚ to scroll down to HI PWR= and press ⬚Enter⬚ to set. Use ⬚Circle⬚ and ⬚Diamond⬚ to adjust. You can choose from 0.1 W, 0.5 W, 1.0 W, 2.0 W and 5.0 W. Press ⬚Enter⬚ to set.

15. Press ⬚Square⬚ repeatedly to escape out of the menus (or just wait).

Lock/unlock radio

While holding ⬚Square⬚, hold ⬚Enter⬚ for one half second, then release both buttons. Repeat this to cycle through:

- KEYS DISABLED/SIDE ENABLED

- KEYS DISABLED/SIDE DISABLED

- KEYS ENABLED/SIDE ENABLED (unlocked).

Release all buttons for one second to set.

Check repeater input frequency

Although talkaround can be programmed using a computer programmer, this is not possible in the field. You will have to program a separate frequency for this. If a frequency has been programmed with talkaround, press ⬚Enter⬚ and then ⬚Circle⬚ to scroll down to SELECT. Press ⬚Enter⬚. Press ⬚Diamond⬚ to scroll up to TKRD=. Press ⬚Enter⬚, then press ⬚Circle⬚ to scroll to ON or OFF. Press ⬚Enter⬚ to set. Press ⬚Square⬚ repeatedly to escape out of the menus. If you are in talkaround mode, the main screen will show TA. Otherwise it will show ⋀⋀⋀ to indicate repeater mode.

Change power in the field

Using a PC programmer, one of the side buttons can be programmed to switch between low and high power. If one is not programmed, you can change the level for an individual channel. See "Program freq in the field" for instructions. The display will show LO or HI to indicate current power level.

Adjust volume

Turn the power/volume knob to adjust volume. The knob has sixteen steps: off, MUTE and then fourteen levels of volume.

Adjust squelch

This is set on a per-channel basis. See "Program freq in the field" for instructions. If a side button has been programmed, you can press it to monitor temporarily, hold it for two seconds to monitor continuously, or hold it for four seconds to adjust squelch. (Use Circle and Diamond to adjust.)

Weird Modes

Buttons don't work

First, check to make sure the side buttons are enabled (see "Lock/Unlock radio" above). If that doesn't work, be aware that many of this radio's buttons and switches are programmable using a PC programmer. Depending on how the radio was configured, the side buttons may or may not work, and may do different things.

Radio has open mic, flashing lights and/or sounds

The radio has an emergency mode. This transmits a signal on P25, but on analog channels it just opens the mic. The radio will also display EMERGENCY. Depending on the emergency mode programmed, there may also be exciting lights and sounds. Turn the radio off and on again to exit.

Useful Information

⚠ **WARNING:** This radio uses an SF (Motorola-style) SMA connector. This requires that the female side of the connector have a center conductor that extends to the end of the connector. Standard SMA connectors do not do this. Be sure your antenna is making contact with the radio's connector.

Standard connector won't work

SF (Motorola-style) SMA will work

In the United States, programming a frequency with NOAA weather radio is a quick way to test if the antenna is making connection.

It is possible to set the radio to display frequency rather than channel name. To do this, press Enter, then use Diamond to scroll to PROGRAM. Press Enter. Ensure that GLOBAL is selected. (If not, use Circle to scroll to it.) Press [Enter]. Press [Circle] to scroll to DISPLY. Press [Enter]. Press [Circle] to scroll between ALPHA (channel name) and NUMBER (frequency). Press [Enter] to set. Press [Square] repeatedly to escape out of the menu system.

The default password is 000000.

When a channel has a square box around it, that means it's in the scan list. Press [Circle] to remove or [Diamond] to add it.

If your radio has the Fire Feature, you can switch to bank 5. Press [Enter], then scroll to SELECT using [Circle]. Press [Enter]. Scroll to ZONE using [Circle]. Press [Enter]. Choose either 1 BANK 01 or 5 BANK 05. Press [Enter]. Press [Square] to escape out of the menu system.

Some radios have a OTAR (over-the-air re-keying) mode that shows up when you first press [Enter]. Use [Circle] to get to the regular menu system.

No reset procedure

This radio does not have a way to reset from the keypad.

⚠ **WARNING:** In at least some configurations, the Thales 25 may permit you to transmit on business or public safety frequencies. Make sure you are in-band when transmitting.

Radio Layout

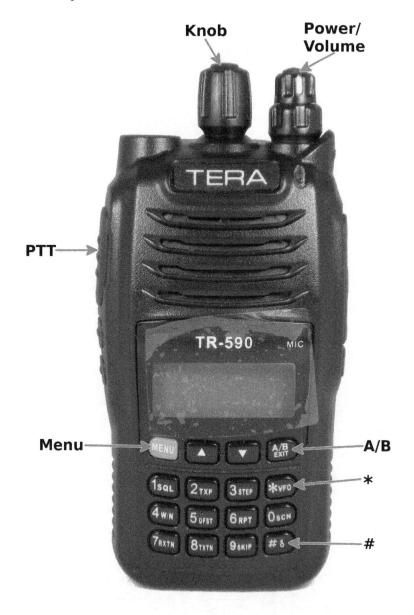

Specs

Receivers Two independent receivers, simultaneous receive
Receives 136–174 MHz, 400–470 MHz
Transmits 136–174 MHz FM @ 5 W, 400–470 MHz FM @ 5 W
Antenna connector SMA **M** on radio; needs SMA **F** antenna
Modes FM
Memory Channels 199
Power 6.6–8.4 V DC per specification, No DC input on radio
Model year 2014

Standard Tasks

Program frequency in the field

1. Press ⬚*⬚ to go to VFO mode if you aren't there already. (You will see Ch nnn in channel mode, where nnn is 001–199.)

2. Decide which memory to use. You can't overwrite—you will have to delete first if that memory has data in it. Press ⬚Menu⬚ ⬚3⬚ ⬚3⬚ ⬚Menu⬚ (Del Ch). Use the knob or keypad to select the channel to delete (001–199). Then press ⬚Menu⬚ ⬚A/B⬚.

3. Enter frequency (144390 for 144.390 MHz).

4. To set transmit tone, press ⬚Menu⬚ ⬚8⬚ ⬚Menu⬚ (Tx Tone) then scroll to the correct CTCSS tone frequency using the center selection knob, press ⬚Menu⬚ ⬚A/B⬚. If you see DCS tones or OFF, press ⬚#⬚ until you see CTCSS tones.

5. To set receive tone, press ⬚Menu⬚ ⬚7⬚ ⬚Menu⬚ (Rx Tone) then scroll to the correct CTCSS tone frequency using the center selection knob, press ⬚Menu⬚ ⬚A/B⬚. If you see DCS tones or OFF, press ⬚#⬚ until you see CTCSS tones.

6. Press ⬚Menu⬚ ⬚6⬚ ⬚Menu⬚ (Rpt Dir) and scroll to the correct repeater shift (+, – or OFF). Press ⬚Menu⬚ ⬚A/B⬚.

7. Press ⬚Menu⬚ ⬚5⬚ ⬚Menu⬚ (Offset) and scroll to the correct repeater offset, then press ⬚Menu⬚ ⬚A/B⬚.

8. Press ⬚Menu⬚ ⬚2⬚ ⬚Menu⬚ (TXPower) and scroll to correct power level (LOW (0.5 W), MID (3 W VHF, 2 W UHF) or HIGH (5 W), then press ⬚Menu⬚ ⬚A/B⬚.

9. Press ⬚Menu⬚ ⬚3⬚ ⬚2⬚ ⬚Menu⬚ (Save Ch). Use the knob or keypad to select the channel to write (001–199) then press ⬚Menu⬚ ⬚A/B⬚.

10. Press ⬚*⬚ to enter channel (memory) mode.

11. Scroll to the channel you just wrote.

Lock/unlock radio

Hold [#] for three seconds to lock/unlock.

Check repeater input frequency

Hold [*] for two seconds. The R indicator will show you're in reverse mode.
Hold [*] again to return to normal mode.

Change power in the field

Press [Menu] [2] [Menu] (TXPower) and scroll to correct power level (LOW
(0.5 W), MID (3 W VHF, 2 W UHF) or HIGH (5 W), then press [Menu] [A/B].

Adjust volume

Rotate the right power/volume knob to adjust volume.

Adjust squelch

Press [Menu] [1] [Menu] (SqlLevel) then scroll to the correct squelch level
(0–9). Then press [Menu] [A/B].

Weird Modes

Keys don't work and display shows FM

You are in FM radio mode. This is typically entered via a side button;
pressing that button again will leave the mode. If that doesn't work, try
[Menu] [1] [9] (FmRadio) or [Menu] [2] [0] (Fm Mon). Press [Menu], then
use the knob to change to OFF. Then press [Menu] [A/B].

Side buttons don't behave as expected

This radio has two buttons on the left hand side below [PTT]. They can be
programmed for a number of different functions.

Can't leave channel (memory) mode

This radio is often shipped with VFO mode turned off. If that's the case,
you will need to modify the programming with a computer / radio programming cable to turn VFO mode on. This cannot be changed from the
front panel.

Useful Information

To switch between monitoring one frequency and monitoring two frequencies, use [Menu] [2] [2] [Menu] (SubDisp) then scroll to either Dual or
Single, then press [Menu] [A/B].

Factory reset

To reset the radio, hold down $\boxed{\text{A/B}}$ while turning the radio on. Use the knob to choose ALL to reset everything. Press $\boxed{\text{Menu}}$.

VFO reset

To reset the radio, hold down $\boxed{\text{A/B}}$ while turning the radio on. Use the knob to choose VFO to reset VFO-related settings only). Press $\boxed{\text{Menu}}$.

⚠ **WARNING:** In at least some configurations, the Tera TR-590 may permit you to transmit on business or public safety frequencies. Make sure you are in-band when transmitting.

Tytera TYT MD-2017

Radio Layout

Power/
Volume

PTT

Up

Down

Trackball

Menu

*

Specs

Receivers Single receiver
Receives 136–174 MHz FM / DMR, 400–480 MHz FM / DMR
Transmits 136–174 MHz FM / DMR @ 5 W , 400–480 MHz FM / DMR @
    5 W
Antenna connector SMA F on radio; needs SMA M antenna
Modes FM, DMR
Memory Channels 3000
Power 7.4 V DC, No DC input on radio
Model year 2017

Standard Tasks

Program frequency in the field

This radio allows you to add new frequencies in the field. Frequencies must be added to a zone. You can add zones in the field as well. Many codeplugs have an "FPP" zone intended for front-panel programming. You can use an existing zone rather than adding a new zone if the zone has room in it.

⚠ **WARNING:** This radio can be configured not to allow front panel programming, or to allow it only with a password. If it's configured to disable FPP, or if you don't know the password, you will not be able to program the radio.

1. Enter programming mode: press [Menu] [Up] [Menu] to select the Utilities menu. Then press [Up] [Menu] to select the menu item Program Radio. Enter the password if required (many code plugs use 99999999 or 00000000 as the eight digit password), then press [Menu].

2. Press [Down] or use the trackball to scroll to menu 6, Add CH. Press [Menu].

3. Make sure item 1, Analog CH is selected. Push [Menu].

4. You will be prompted with Enter CH Name. You can use the trackball to adjust the cursor position, and the keypad to enter a name. Press [Menu].

5. You will be prompted with Rx Frequency. Use the trackball to back up and then the keypad to enter (144390 for 144.390 MHz). Press [Menu].

6. You will be prompted with Tx Frequency. Use the trackball to back up and then the keypad to enter (144390 for 144.390 MHz). Note that you set receive and transmit frequency rather than an offset. Press [Menu]. The channel is added.

7. To add the channel to a new zone, press [Menu] and select Zone with the trackball. Press [Menu].

8. Select New Zone with the trackball. You will be prompted for a zone name. Use the trackball to adjust cursor position and the keypad to enter a zone name. Press [Menu].

9. To select the new zone, press [Menu] and use the trackball to select [Zone]. Press [Menu].

10. Use the trackball to choose ZoneList. Press [Menu].

11. Use the trackball to choose Add CH. Press [Menu].

12. Select CH A and press [Menu]. You will see Add Channel Successful. Your channel is now added to the zone.

13. Turn the zone on. Press [Menu] then select Zone, press [Menu] and select ZoneList. Press [Menu]. Scroll to your zone and press [Menu]. Select menu item 1, On. Press [Menu].

14. Your channel is now set in your zone. Make sure your channel is selected, then press [Menu]. Use the trackball to scroll to Utilities and press [Menu].

15. Use the trackball to scroll to Program Radio and press [Menu].

16. Scroll to Edit Channel and press [Menu].

17. You can now scroll to CTC/DCS and press [Menu].

18. Scroll to T CTC and press [Menu].

19. Use the trackball up/down to select the tone you want. Press [Menu] when done.

20. Press [Menu] then [Up] [Menu] to select Utilities. Press [Menu] to choose Radio Settings. Press [Down] twice to select Power. Press [Menu] to confirm. Use [Up] or [Down] to select the power level you want (High is 5 W, Low is 1 W), then press [Menu] to set.

Lock/unlock radio

To lock the radio, press [Menu] and go to the Utilities menu. Press [Menu]. Scroll to menu 7, Keypad Lock. Press [Menu]. Use the trackball to choose from Manual (no lock), or a time to lock (5, 10 or 15 seconds). Press [Menu]. Wait for the radio to lock.

To unlock the radio, press [Menu] [*].

Check repeater input frequency

You can't switch to the repeater input frequency in the field. You will have to program a separate memory for this.

Change power in the field

Press ⌈Menu⌉ then ⌈Up⌉ ⌈Menu⌉ to select Utilities. Press ⌈Menu⌉ to choose Radio Settings. Press ⌈Down⌉ twice to select Power. Press ⌈Menu⌉ to confirm. Use ⌈Up⌉ or ⌈Down⌉ to select the power level you want (High is 5 W, Low is 1 W), then press ⌈Menu⌉ to set.

Adjust volume

Use the Power/Volume knob to adjust the volume.

Adjust squelch

Press ⌈Menu⌉ then ⌈Up⌉ then ⌈Menu⌉ to select Utilities. Then press ⌈Menu⌉ to choose Radio Settings. Press ⌈Down⌉ four times to select Squelch. Press ⌈Menu⌉ to confirm. Use ⌈Up⌉ or ⌈Down⌉ to select the squelch you want (Tight or Normal). Press ⌈Menu⌉ to set.

Weird Modes

Radio doesn't transmit

This radio has a transmit inhibit feature. To disable it, press ⌈Left Menu⌉ then ⌈Up⌉ then ⌈Menu⌉ to select Utilities. Then press ⌈Menu⌉ to choose Radio Settings. Press ⌈Down⌉ to select Tones/Alerts, then press ⌈Menu⌉ to confirm. Press ⌈Down⌉ to select Talk Permit, and press ⌈Menu⌉. Use ⌈Up⌉ or ⌈Down⌉ to choose Turn On (user can transmit) or Turn Off (user can't transmit). Press ⌈Menu⌉ to set.

Radio stops receiving and/or transmitting after receiving a signal

This radio has a remote kill feature that can prevent the radio from transmitting or receiving when it receives a properly-formatted transmission. Most of the time, this requires programming software to revive the radio. (There is a remote activation feature, but odds are good you won't be able to send that.)

Useful Information

The radio's memories are divided into zones (similar to banks), with a maximum of sixteen memories per bank. To change zones, from the root menu press ⌈Menu⌉ then ⌈Down⌉ to select Zone. Press ⌈Menu⌉ to confirm. Use the trackball to select ZoneList and press ⌈Menu⌉. Select your zone and press ⌈Menu⌉. Select On and press ⌈Menu⌉ to confirm. To change memories within a zone, use the trackball.

The ⌐Menu⌐ button on the radio has a top function and a bottom function. You need to press the bottom of the button to get the ⌐Menu⌐ function.

This radio has an unusual antenna connector, which is wider than normal (although still SMA).

MD-2017 antenna

On at least some versions of the radio, pressure on the antenna could snap the SMA connector, requiring replacement. Aftermarket antennas may put additional stress on the radio's SMA connector.

The radio menu system has a 10-second timeout. You will need to act quickly once you have entered the menu.

This radio has the ability to record received audio.

Although this radio is dual-band, only one band can be active at a time.

This radio has frequent firmware updates (and sometimes has third-party firmware loaded as well). These instructions are for stock firmware V003.033.

No reset procedure

The radio does not have a way to reset it without programming a new codeplug. (The radio does reset when a new codeplug is uploaded.)

⚠ **WARNING:** In at least some configurations, the Tytera TYT MD-2017 may permit you to transmit on business or public safety frequencies. Make sure you are in-band when transmitting.

Tytera TYT MD-380

Radio Layout

Left Knob

Power/ Volume

PTT

Up

Down

Left Menu

Right Menu

*

Specs

Receivers Single receiver
Receives One of 136–174 MHz FM/DMR or 400–480 MHz FM/DMR
Transmits One of 136–174 MHz FM / DMR @ 5 W or 400–480 MHz FM / DMR @ 5 W
Antenna connector SMA F on radio; needs SMA M antenna
Modes FM, DMR
Memory Channels 1000
Power No DC input on radio; 12.5 V DC, 5.5 mm OD, 2.1 mm ID, center positive on charger base
Model year 2015

Standard Tasks

Program frequency in the field

This radio does not allow you to add new frequencies in the field. However, you can modify existing frequencies that have been programmed. Many codeplugs have a "FPP" zone intended for this purpose.

⚠ **WARNING:** This radio can be configured not to allow front panel programming, or to allow it only with a password. If it's configured to disable FPP, or if you don't know the password, you will not be able to program the radio.

1. Set the radio to the memory that you want to overwrite. Note that you can't convert an FM memory into a DMR memory or vice versa, and you can't change talkgroup. (Note that you may need to change zones; see "Useful Info" below.)

2. Enter programming mode: press ⌞Left Menu⌟ ⌞Up⌟ ⌞Left Menu⌟ to select the Utilities menu. Then press ⌞Up⌟ ⌞Left Menu⌟ to select the menu item Program Radio. Enter the password (many code plugs use 99999999 or 00000000 as the eight digit password), then press ⌞Left Menu⌟.

3. Press ⌞Left Menu⌟ ⌞Left Menu⌟ to edit receive frequency.

4. Press ⌞Up⌟ repeatedly to delete previous frequency.

5. Enter the desired frequency on the keypad (445925 for 445.925 MHz). Press ⌞Left Menu⌟ to set.

6. Press ⌞Down⌟ ⌞Left Menu⌟ ⌞Left Menu⌟ to edit transmit frequency.

7. Press ⌞Up⌟ repeatedly to delete previous frequency.

8. Enter the desired frequency on the keypad (445925 for 445.925 MHz). Press ⌞Left Menu⌟ to set.

9. Press ⌞Down⌟ ⌞Down⌟ ⌞Down⌟ ⌞Left Menu⌟ to enter the CTC/DCS menu. (If you can't get into that menu, you're on a DMR memory and not an analog memory.)

10. Press [Left Menu] to go into the R CTC (receive CTCSS tone) menu. Press [Left Menu] to change, use [Up] and [Down] to select, then press [Left Menu] to set. (You can change R DCS instead if you use DCS tones.)

11. Press [Down] [Down] [Left Menu] to set T CTC (transmit CTCSS tone). Press [Left Menu] to change, use [Up] and [Down] to select, then press [Left Menu] to set. (You can change R DCS instead if you use DCS tones.)

12. Press [Right Menu] repeatedly to exit the menu system.

13. Press [Left Menu] then [Up] then [Left Menu] to select Utilities. Press [Left Menu] to choose Radio Settings. Press [Down] twice to select Power. Press [Left Menu] to confirm. Use [Up] or [Down] to select the power level you want (High is 5 W, Low is 1 W), then press [Left Menu] to set.

Lock/unlock radio

Hold [*] for two seconds to lock. Press [Left Menu] [*] to unlock when locked. (Note that the radio can be configured to auto-lock by going into the Radio Settings menu with Keypad Lock.)

Check repeater input frequency

You can't switch to the repeater input frequency in the field. You will have to program a separate memory for this.

Change power in the field

Press [Left Menu] then [Up] then [Left Menu] to select Utilities. Then press [Left Menu] to choose Radio Settings. Press [Down] twice to select Power. Press [Left Menu] to confirm. Use [Up] or [Down] to select the power level you want (High is 5 W, Low is 1 W), then press [Left Menu] to set.

Adjust volume

Use the Power/Volume knob to adjust the volume.

Adjust squelch

Press [Left Menu] then [Up] then [Left Menu] to select Utilities. Then press [Left Menu] to choose Radio Settings. Press [Down] four times to select Squelch. Press [Left Menu] to confirm. Use [Up] or [Down] to select the squelch you want (Tight or Normal). Press [Left Menu] to set.

Weird Modes

Radio doesn't transmit

This radio has a transmit inhibit feature. To disable it, press [Left Menu] then [Up] then [Left Menu] to select Utilities. Then press [Left Menu] to choose Radio Settings. Press [Down] to select Tones/Alerts, then press [Left Menu] to confirm. Press [Down] to select Talk Permit, and press [Left Menu]. Use [Up] or [Down] to choose Turn On (user can transmit) or Turn Off (user can't transmit). Press [Left Menu] to set.

Radio stops receiving and/or transmitting after receiving a signal

This radio has a remote kill feature that can prevent the radio from transmitting or receiving when it receives a properly-formatted transmission. Most of the time, this requires programming software to revive the radio. (There is a remote activation feature, but odds are good you won't be able to send that.)

Radio menus are in Chinese

To reset the radio to English, first press [Left Menu], [Up], [Left Menu], then [Left Menu]. Press [Down] seven times (until you are on 8) then press [Left Menu]. Press [Up] to select English and press [Left Menu] to set.

Useful Information

The radio's memories are divided into zones (similar to banks), with a maximum of sixteen memories per bank. To change zones, from the root menu press [Left Menu] then [Down] to select Zone. Press [Left Menu] to confirm. Press [Up] and [Down] to select a zone, then press [Left Menu] to confirm. To change memories within a zone, use the left knob.

The radio menu system has a ten-second timeout. You will need to act quickly once you have entered the menu. In particular, you will need to re-enter the programming password if you time out.

No reset procedure

The radio does not have a way to reset it without programming a new codeplug. (The radio does reset when a new codeplug is uploaded.)

⚠ **WARNING:** In at least some configurations, the Tytera TYT MD-380 may permit you to transmit on business or public safety frequencies. Make sure you are in-band when transmitting.

Tytera TYT TH-F8

Radio Layout

Specs

Receivers Single receiver, dual watch (first to break squelch wins)
Receives Single band: one of 136–174 MHz FM or 245–246 MHz FM or
    400–470 MHz FM or 350–390 MHz FM or 465–520 MHz FM
Transmits Single band: one of 136–174 MHz FM @ 5 W or 245–246 MHz
    FM @ 5 W or 400–470 MHz FM @ 4 W or 350–390 MHz FM @ 4 W or
    465–520 MHz FM @ 4 W
Antenna connector SMA F on radio; needs SMA M antenna
Modes FM
Memory Channels 128
Power No DC input on radio; 12 V DC, 5.5 mm OD, 2.1 mm ID, **center
    negative** on charger base
Model year 2010

Standard Tasks

Program frequency in the field

1. Press [Exit] to go to VFO mode if you aren't there already. (Channel
   number is not shown in VFO mode.)

2. Depending on the frequency you want to program, it may be neces-
   sary to adjust the step beforehand. To do this, Press [F] [2] [9] [F].
   The display will show STEP - 29 and the current value. Use [Up] and
   [Down] to change, then press [F] [Exit] [Exit] to save.

3. Enter the desired frequency on the keypad (144390 for 144.390 MHz).

4. To set the tone, press [F] [2] [7] [F]. The display will show T-CDC-
   - 27. Then press [*] until you see 67.0 (for CTCSS) or D023N (for
   DCS). Use [Up] and [Down] to set the value to your choice. Note that
   menu 27 sets the transmit tone; you can use menu 26 (R-CDC - 26)
   for receive tone and menu 25 (C-CDC - 25) for both transmit and
   receive tone. Once you have set the tone, press [F] [Exit] [Exit] to
   save.

5. Press [F] [2] [3] [F] to set repeater offset (display will show OFFSET-
   23 and the current value). Press [Up] and [Down] to adjust, or enter
   the offset directly on the number pad. (00600 is the standard 600
   kHz offset, which is displayed as 0.600.) Press [F] [Exit] [Exit] to
   save.

6. Press [F] [2] [8] [F] to set repeater shift (display will show S-D - 28
   and the current value). Press [Up] and [Down] to adjust among OFF, -
   and +. Press [F] [Exit] [Exit] to save.

7. Press [F] [4] [F] to set power level (display will show POW - 04 and
   the current value). Press [Up] and [Down] to adjust between HIGH (5 W
   VHF or 4 W UHF) and LOW (1 W VHF or 0.5 W UHF). Press [F] [Exit]
   [Exit] to save.

8. Press [F] [Exit] to write to a memory. Use [Up] and [Down] to select the memory to write (memory will flash if it already has data in it). Press [Exit] to store. You will automatically enter memory mode after writing the memory.

Lock/unlock radio

Hold [*] for two seconds to lock/unlock.

Check repeater input frequency

Hold [#] for two seconds to switch to reverse frequency. (Display will show R when in reverse mode.) Hold [#] for two seconds to leave reverse frequency mode.

Change power in the field

Press [F] [4] [F] to set power level (display will show POW - 04 and the current value). Press [Up] and [Down] to adjust between HIGH (5 W VHF or 4 W UHF) and LOW (1 W VHF or 0.5 W UHF). Press [F] [Exit] [Exit] to save.

Adjust volume

Use the Power/Volume knob to adjust the volume.

Adjust squelch

Press [F] [5] [F] to set power level (display will show SQL - 05 and the current value). Press [Up] and [Down] to adjust from 0–9 (default is 5). Press [F] [Exit] [Exit] to save.

Weird Modes

Transmit audio is distorted

This radio has a built-in scrambler which may not be legal for amateur radio use. To check, press Press [F] [3] [4] [F]. If the radio displays APRO - 34 then it has the scrambler. Use [Up] and [Down] to set it to OFF, then press [F] [Exit] [Exit] to save.

Radio stops receiving and/or transmitting after receiving a signal

This radio has a remote kill feature that can prevent the radio from transmitting or receiving when it receives a properly-formatted transmission. Most of the time, this requires programming software to revive the radio.

Radio doesn't enter VFO mode

The radio has a "Channel Mode" that prevents direct frequency entry. To enable/disable this mode, hold down [Up] while turning the radio on.

Useful Information

The password for the AGING ? reset is often 5858.

Some users have reported problems after using the programming software. The radio goes into a mode where its output power is limited to 1 W. The dealer programming software may be able to reset this.

Factory reset

Hold down [F] while turning the radio on, then press [F]. The display will show VFO ?. Press [Up] / [Down] to select FULL ? (resets everything). (If you decide you don't want to reset, turn the radio off and back on.) Press [F] to perform the reset.

VFO reset

Hold down [F] while turning the radio on, then press [F]. The display will show VFO ?. Press [Up] / [Down] to select FULL ? (resets everything). (If you decide you don't want to reset, turn the radio off and back on.) Press [F] to perform the reset.

Aging Reset

This radio has a third kind of reset. The meaning of this is not known, but it may have something to do with allowing / preventing out-of-band transmissions. Hold down [F] while turning the radio on, then press [F]. The display will show VFO ?. Press [Up] / [Down] to select AGING ?. (If you decide you don't want to reset, turn the radio off and back on.) Press [F] to perform the reset.

⚠ **WARNING:** In at least some configurations, the Tytera TYT TH-F8 may permit you to transmit on business or public safety frequencies. Make sure you are in-band when transmitting.

Wouxun KG-UVD1P, -UV2D, -UV3D

Radio Layout

Knob

Power/ Volume

PTT

TDR

Menu

Exit

Scan *

#

KG-UV2D model

Specs

Receivers Single receiver, dual watch (first to break squelch wins)
Receives 76–108 MHz WFM, 136–174 MHz FM and 420–520 MHz FM, 136–174 MHz and 350–470 MHz FM, or 136–174 MHz and 216–280 MHz FM (depending on model)
Transmits 136–174 MHz @ 5 W FM and 420–520 MHz @ 4 W FM, 136–174 MHz @ 5 W FM and 350–470 MHz @ 4 W FM, or 136–174 MHz @ 5 W FM and 216–280 MHz @ 5 W FM (depending on model)
Antenna connector SMA **M** on radio; needs SMA **F** antenna
Modes FM
Memory Channels 128
Power No DC input on radio; charger has "M"-style barrel (5.5 mm OD, 2.1 mm ID) with 12 V DC center positive
Model year 2010

Standard Tasks

Program frequency in the field

1. Decide which memory to use. You can't overwrite—you will have to delete first if that memory has data in it. Press [Menu] [2] [8] [Menu] (DEL-CH) *XXX* where *XXX* is the channel (001–128). Then press [Menu].

2. Press [Menu] then [TDR] to go to VFO mode if you aren't there already.

3. Enter frequency (144390 for 144.390 MHz).

4. Press [Menu] [1] [6] [Menu] (T-CTC) then scroll to the correct CTCSS tone frequency (or off) using the center selection knob, press [Menu] [Exit].

5. Press [Menu] [2] [4] [Menu] (SFT-D) and scroll to the correct repeater shift, press [Menu] [Exit].

6. Press [Menu] [2] [3] [Menu] (OFFSET) and scroll to the correct repeater offset, then press [Menu] [Exit].

7. Press [Menu] [4] [Menu] (TXP) and scroll to correct power level, then press [Menu] [Exit].

8. Press [Menu] [2] [7] [Menu] (MEM-CH) and enter channel to write *XXX* (001–128) then press [Menu] [Exit].

9. Press [Menu] [TDR] to enter channel (memory) mode.

10. Scroll to the channel you just wrote.

Lock/unlock radio

Hold [#] for five seconds to lock/unlock.

Check repeater input frequency

Press [Scan *] for less than one second and release. R indicator will show you're in reverse mode. Press and release [Scan *] again to return to normal mode.

Change power in the field

Press [Menu] [4] [Menu] and scroll to correct power level (LOW or HIGH), press [Menu] [Exit].

Adjust volume

Rotate the right power/volume knob to adjust volume.

Adjust squelch

Press [Menu] [2] [Menu] (SQL-LE) then scroll to the correct squelch level, press [Menu] [Exit].

Weird Modes

Radio displays countdown when locking

This is normal for some models; ignore it.

Entering frequency "doesn't take"

You are probably in channel (memory) mode with frequency display turned on, instead of VFO mode. Switch to VFO mode.

Useful Information

Menu numbers are for the KG-UV2D. Other models may have them in different positions.

If you wait too long after pressing [Menu], the menu mode will time out. You need to press quickly.

This radio comes in three models: two for 2m / 70cm and one for 2m / 1.25m.

This radio has a "dealer mode" that allows overwriting of memories. To enter this, hold down [8] while turning the radio on, then enter 268160. This switches channel display (menu 21, CH-MDF)to CH instead of NAME or FREQ, so you will have to set it back using the menu. Dealer mode will be inactivated when you turn off the radio.

Factory reset

To reset the radio, press [Menu] [2] [9] [Menu]. Press [Down] to choose ALL. Press [Menu]. You will see SURE? in the display. (If the device has a password set, you will see - - - - - - and you must enter the correct six-digit password to continue.) Push [Menu] to reset.

VFO reset

To reset the radio, press [Menu] [2] [9] [Menu]. Press [Down] to choose VFO if it isn't already selected. Press [Menu]. You will see SURE? in the display. (If the device has a password set, you will see - - - - - - and you must enter the correct 6-digit password to continue.) Push [Menu] to reset.

⚠ **WARNING:** In at least some configurations, the Wouxun KG-UVD1P, KG-UV2D and KG-UV3D may permit you to transmit on business or public safety frequencies. Make sure you are in-band when transmitting.

Wouxun KG-UV3X Pro

Radio Layout

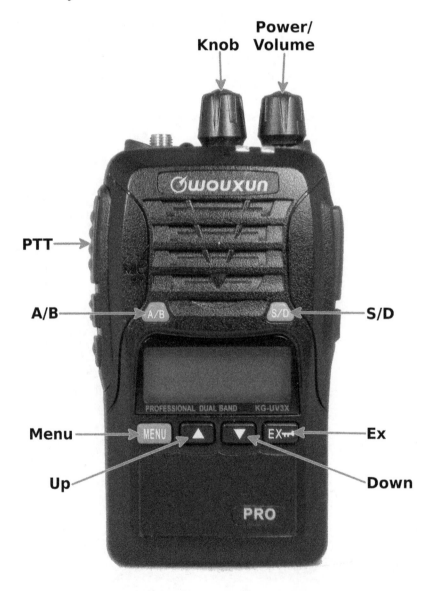

Specs

Receivers Single receiver, dual watch (first to break squelch wins)
Receives 76–108 MHz WFM, 136–174 MHz FM and 375–512 MHz FM
Transmits 136–174 MHz @ 5 W FM and 375–512 MHz @ 4 W FM
Antenna connector SMA F on radio; needs SMA M antenna
Modes FM
Memory Channels 128
Power No DC input on radio; charger has "M"-style barrel (5.5 mm OD, 2.1 mm ID) with 12 V DC center positive
Model year 2012

Standard Tasks

Program frequency in the field

1. Decide which memory to use. You can't overwrite—you will have to delete first if that memory has data in it. Use the knob to select the channel to delete. Press Menu then use knob to scroll to menu 28, DEL-CH. Press Menu Menu to delete. Press Ex.

2. Press Menu then S/D to go to VFO mode if you aren't there already. (You may see a channel number if you aren't in VFO mode.)

3. Use knob to adjust frequency.

4. To set tone frequency, press Menu then use the knob to select menu 16, T-CTC. Press Menu then use the knob to select the frequency. Press Menu Ex. You can also adjust R-CTC (menu 15), R-DCS (menu 17) and T-DCS (menu 18) if desired.

5. To set shift, press Menu then use the knob to select menu 24, SFT-D. Press Menu to alter, and use the knob to select from OFF, + and -. Press Menu Ex.

6. To set offset, press Menu then use the knob to select menu 23, OFFSET. Press Menu to alter, and use the knob to adjust. Press Menu Ex.

7. To set transmit power, press Menu then use the knob to select menu 4, TXP. Press Menu to alter, and use the knob to select from HIGH (5 W VHF, 4 W UHF) and LOW (1 W). Press Menu Ex.

8. To write to memory, press Menu then use the knob to select menu 27, MEM-CH. Press Menu then use the knob to select the desired channel. You will not be able to write into any channel that already has something stored in it. Press Menu Ex.

9. Press Menu S/D to enter channel (memory) mode.

10. Scroll to the channel you just wrote.

Lock/unlock radio

Hold [Ex] for two seconds to lock/unlock.

Check repeater input frequency

There is no way to check the input except to program another memory.

Change power in the field

press [Menu] then use the knob to select menu 4, TXP. Press [Menu] to alter, and use the knob to select from HIGH (5 W VHF, 4 W UHF) and LOW (1 W). Press [Menu] [Ex]

Adjust volume

Rotate the right power/volume knob to adjust volume.

Adjust squelch

ress [Menu] then use the knob to select menu 2, SQL-LE. Press [Menu] then use the knob to select the squelch level (0–9). Press [Menu] [Ex].

Weird Modes

Can't switch to VFO or memory (channel) mode

This radio is shipped as a Part 90 (commercial) radio. You will first have to enable front panel programming using a computer, programming cable and software before you can do anything from the keypad. In addition, a factory reset of the radio will disable front panel programming.

Entering frequency "doesn't take"

You are probably in channel (memory) mode with frequency display turned on, instead of VFO mode. Switch to VFO mode.

Useful Information

If you wait too long after pressing [Menu], the menu mode will time out. You need to press quickly.

This radio often ships with mode switch password 123456 and reset password 654321. If the password is 000000, you won't be prompted.

You can use [Up] / [Down] in place of the knob.

Factory reset

To reset the radio, press [Menu] then use the knob to select menu 29, RESET. Press [Menu]. Use the knob to choose ALL. Press [Menu]. You will see SURE? in the display. (If the device has a password set, you will see

- - - - - - and you must enter the correct 6-digit password to continue.) Push ⟦Menu⟧ to reset.

⚠ **WARNING:** A factory reset will require you to use a computer, programming cable and software to re-activate front panel programming.

VFO reset

To reset the radio, press ⟦Menu⟧ then use the knob to select menu 29, RESET. Press ⟦Menu⟧. Use the knob to choose ALL. Press ⟦Menu⟧. You will see SURE? in the display. (If the device has a password set, you will see - - - - - - and you must enter the correct 6-digit password to continue.) Push ⟦Menu⟧ to reset.

⚠ **WARNING:** In at least some configurations, the Wouxun KG-UV3X Pro may permit you to transmit on business or public safety frequencies. Make sure you are in-band when transmitting.

Wouxun KG-UV6D, KG-UV6X

Radio Layout

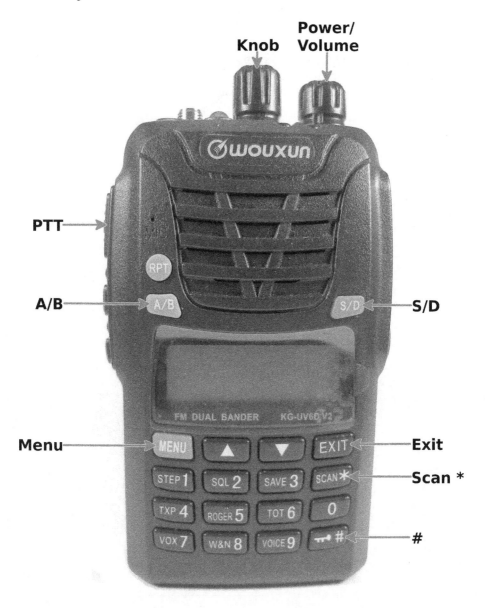

Specs

Receivers Single receiver, dual watch (first to break squelch wins)

Receives 76–108 MHz WFM, 136–174 MHz FM and 420–520 MHz FM, 136–174 MHz and 350–470 MHz FM, or 136–174 MHz and 216–280 MHz FM (depending on model)

Transmits 136–174 MHz @ 5 W FM and 420–520 MHz @ 4 W FM, 136–174 MHz @ 5 W FM and 350–470 MHz @ 4 W FM, or 136–174 MHz @ 5 W FM and 216–280 MHz @ 5 W FM (depending on model)

Antenna connector Most ship with SMA F on radio; need SMA M antenna. A few have shipped with the less-common SMA **M** on radio and need an SMA F antenna.

Modes FM

Memory Channels 199

Power No DC input on radio; charger has "M"-style barrel (5.5 mm OD, 2.1 mm ID) with 12 V DC center positive

Model year 2012

Standard Tasks

Program frequency in the field

1. Decide which memory to use. You can't overwrite—you will have to delete first if that memory has data in it. Press [Menu] [2] [9] [Menu] (DEL-CH) *XXX* where *XXX* is the channel (001–199). Then press [Menu].

2. Press [Menu] then [S/D] to go to VFO mode if you aren't there already.

3. Enter frequency (144390 for 144.390 MHz).

4. Press [Menu] [1] [6] [Menu] (T-CTC) then scroll to the correct CTCSS tone frequency (or off), press [Menu] [Exit].

5. Press [Menu] [2] [5] [Menu] (SFT-D) and scroll to the correct repeater shift, press [Menu] [Exit].

6. Press [Menu] [2] [4] [Menu] (OFFSET) and scroll to the correct repeater offset, then press [Menu] [Exit].

7. Press [Menu] [4] [Menu] (TXP) and scroll to correct power level, then press [Menu] [Exit].

8. Press [Menu] [2] [8] [Menu] (MEM-CH) and enter channel to write *XXX* (001–199) then press [Menu] [Exit].

9. Press [Menu] [A/B] to enter channel (memory) mode.

10. Scroll to the channel you just wrote.

Lock/unlock radio

Hold [#] for two seconds to lock/unlock.

Check repeater input frequency

Press [Scan *] for less than one second and release. R indicator will show you're in reverse mode. Press and release [Scan *] again to return to normal mode.

Change power in the field

Press [Menu] [4] [Menu] and scroll to correct power level (LOW or HIGH), press [Menu] [Exit].

Adjust volume

Rotate the right power/volume knob to adjust volume.

Adjust squelch

Press [Menu] [2] [Menu] (SQL-LE) then scroll to the correct squelch level, press [Menu] [Exit].

Weird Modes

Radio displays countdown when locking

This is normal for some models; ignore it.

Entering frequency "doesn't take"

You are probably in channel (memory) mode with frequency display turned on, instead of VFO mode. Switch to VFO mode.

Useful Information

If you wait too long after pressing [Menu], the menu mode will time out. You need to press quickly.

This radio comes in three models: two for 2m / 70cm and one for 2m / 1.25m.

Factory reset

To reset the radio, press [Menu] [3] [0] [Menu]. Press [Down] to choose ALL. Press [Menu]. You will see SURE? in the display. (If the device has a password set, you will see - - - - - - and you must enter the correct six-digit password to continue.) Push [Menu] to reset.

VFO reset

To reset the radio, press [Menu] [3] [0] [Menu]. Press [Down] to choose VFO if it isn't already selected. Press [Menu]. You will see SURE? in the display. (If the device has a password set, you will see - - - - - - and you must enter the correct 6-digit password to continue.) Push Menu to reset.

⚠ **WARNING:** In at least some configurations, the Wouxun KG-UV6D and KG-UV6X may permit you to transmit on business or public safety frequencies. Make sure you are in-band when transmitting.

Wouxun KG-UV8D

Radio Layout

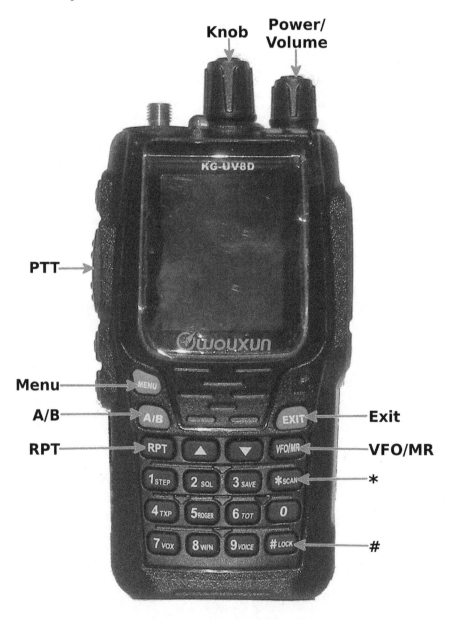

Specs

Receivers Two independent receivers, simultaneous receive

Receives 76–108 MHz WFM, 136–174 MHz FM and 420–520 MHz FM, or 136–174 MHz and 400–480 MHz FM (depending on model)

Transmits 136–174 MHz @ 5 W FM and 420–520 MHz @ 4 W FM, or 136–174 MHz @ 5 W FM and 400–480 MHz @ 4 W FM (depending on model)

Antenna connector SMA **M** on radio; needs SMA **F** antenna

Modes FM

Memory Channels 999

Power No DC input on radio; charger has "M"-style barrel (5.5 mm OD, 2.1 mm ID) with 12 V DC center positive

Model year 2014

Standard Tasks

Program frequency in the field

1. Press [VFO/MR] to go to VFO mode if you aren't there already. You may have to press it up to three times to cycle through different channel display modes. Avoid anything that looks like it has a channel number.

2. Enter frequency (144390 for 144.390 MHz).

3. Press [Menu] [1] [6] [Menu] (T-CTC) then scroll to the correct CTCSS tone frequency (or off), press [Menu] [Exit]. You can also use R-CTC 15, R-DCS 17 and T-DCS 18 depending on what you want to send/receive.

4. Press [Menu] [2] [4] [Menu] (SFT-D) and scroll to the correct repeater shift, press [Menu] [Exit].

5. Press [Menu] [2] [3] [Menu] (OFFSET) and scroll to the correct repeater offset or enter with the keypad (include leading zeros, 005000 for 5 MHz, 000600 for 600 kHz), then press [Menu] [Exit].

6. Press [Menu] [4] [Menu] and scroll to correct power level—LOW (1 W) or HIGH (5 W VHF, 4 W UHF)—then press [Menu] [Exit].

7. Press [Menu] [2] [7] [Menu] (MEM-CH) and enter channel to write *XXX* (001–999) then press [Menu] [Exit].

8. Press [VFO/MR] to enter channel (memory) mode. You may need to press it up to three times to select the display mode you want.

9. Scroll to the channel you just wrote.

Lock/unlock radio

Hold [#] for two seconds to lock/unlock.

Check repeater input frequency

Press [*] for less than one second and release. R indicator will show you're in reverse mode. Press and release [*] again to return to normal mode.

Change power in the field

Press [Menu] [4] [Menu] and scroll to correct power level—LOW (1 W) or HIGH (5 W VHF, 4 W UHF)—then press [Menu] [Exit].

Adjust volume

Rotate the right power/volume knob to adjust volume.

Adjust squelch

Press [Menu] [2] [Menu] (SQL) then scroll to the correct squelch level, press [Menu] [Exit].

Weird Modes

Radio displays circular arrows in upper left

This radio has crossband repeat capability. It is enabled when the circular arrows are displayed in the upper left. Hold [RPT] for two seconds to enable/disable.

You can set the way crossband works with menu 37, RPT-SET. Use X-DIRPT to *receive* on MAIN and transimt on sub-band. Use X-TWRPT for full two-way crossband.

Radio has roger beep

This radio can have a roger beep turned on before, after, or before and after transmission. Use menu 5 ROGER to turn it off.

Entering frequency "doesn't take"

You are probably in channel (memory) mode with frequency display turned on, instead of VFO mode. Switch to VFO mode.

Useful Information

This radio comes in two models, with different ranges for UHF (400–480 or 420–520).

The KG-UV8E also includes 220 MHz.

Factory reset

To reset the radio, press [Menu] [5] [1] [Menu]. Press [Down] to choose ALL. Press [Menu]. You will see SURE? in the display. (If the device has a password set, you will see - - - - - - and you must enter the correct six-digit password to continue.) Push [Menu] to reset.

VFO reset

To reset the radio, press [Menu] [5] [1] [Menu]. Press [Down] to choose VFO if it isn't already selected. Press [Menu]. You will see SURE? in the display. (If the device has a password set, you will see - - - - - - and you must enter the correct 6-digit password to continue.) Push [Menu] to reset.

⚠ **WARNING:** In at least some configurations, the Wouxun KG-UV8D may permit you to transmit on business or public safety frequencies. Make sure you are in-band when transmitting.

Wouxun KG-UV9D

Radio Layout

Knob
Power/Volume
PTT
TDR/VM
Menu
Exit
*
Down
#

Plus version pictured

Specs

Receivers Two independent receivers, simultaneous receive

Receives 76–108 MHz WFM (B only), 108–136 MHz AM (A only), 136–174 MHz FM, 230–250 MHz FM (A only), 350–400 MHz FM (A only) 400–512 MHz FM, 700–985 MHz FM (A only)

Transmits 136–174 MHz @ 5 W FM and 400–512 MHz @ 4 W FM

Antenna connector SMA F on radio; needs SMA M antenna

Modes FM

Memory Channels 999

Power No DC input on radio; charger has "M"-style barrel (5.5 mm OD, 2.1 mm ID) with 12 V DC center positive

Model year 2015

Standard Tasks

Program frequency in the field

1. Hold ⌈TDR/VM⌉ to go to VFO mode if you aren't there already. You may have to press it up to three times to cycle through different channel display modes. Avoid anything that looks like it has a channel number.

2. Enter frequency (144390 for 144.390 MHz).

3. Press ⌈Menu⌉⌈1⌉⌈7⌉⌈Menu⌉ (T-CTC) then scroll to the correct CTCSS tone frequency (or off), press ⌈Menu⌉⌈Exit⌉. You can also use R-CTC 16, R-DCS 18 and T-DCS 19 depending on what you want to send/receive.

4. Press ⌈Menu⌉⌈6⌉⌈Menu⌉ (SFT-D) and scroll to the correct repeater shift, press ⌈Menu⌉⌈Exit⌉.

5. Press ⌈Menu⌉⌈2⌉⌈8⌉⌈Menu⌉ (OFFSET) and scroll to the correct repeater offset or enter with the keypad (include leading zeros, 005000 for 5 MHz, 000600 for 600 kHz), then press ⌈Menu⌉⌈Exit⌉.

6. Press ⌈Menu⌉⌈5⌉⌈Menu⌉ and scroll to correct power level—LOW (1 W), MIDDLE (2 W) or HIGH (5 W VHF, 4 W UHF)—then press ⌈Menu⌉⌈Exit⌉.

7. Press ⌈Menu⌉⌈3⌉⌈0⌉⌈Menu⌉ (MEM-CH) and enter channel to write *XXX* (001–999) then press ⌈Menu⌉⌈Exit⌉.

8. Hold ⌈TDR/VM⌉ to enter channel (memory) mode. You may need to press it up to three times to select the display mode you want.

9. Scroll to the channel you just wrote.

Lock/unlock radio

Hold ⌈#⌉ for two seconds to lock/unlock.

Check repeater input frequency

Press ⌊ * ⌋ for less than one second and release. R indicator will show you're in reverse mode. Press and release ⌊ * ⌋ again to return to normal mode.

Change power in the field

Press ⌊Menu⌋⌊ 5 ⌋⌊Menu⌋ and scroll to correct power level—LOW (1 W), MIDDLE (2 W) or HIGH (5 W VHF, 4 W UHF)—then press ⌊Menu⌋⌊Exit⌋.

Adjust volume

Rotate the right power/volume knob to adjust volume.

Adjust squelch

Press ⌊Menu⌋⌊ 8 ⌋⌊Menu⌋ (SQL-LE) then scroll to the correct squelch level, press ⌊Menu⌋⌊Exit⌋.

Weird Modes

Radio transmits on wrong frequency

The plus version of this radio has crossband repeat capability. Set menu 43, TYPE-SET to TALKIE to disable.

Radio has roger beep

This radio can have a roger beep turned on before, after, or before and after transmission. Use menu 9 ROGER to turn it off.

Entering frequency "doesn't take"

You are probably in channel (memory) mode with frequency display turned on, instead of VFO mode. Switch to VFO mode.

Useful Information

There are two models, the KG-UV9D and the KG-UV9D+. The plus version has 61 menu entries rather than the 55 of the non-plus version, and includes crossband repeat.

Factory reset

To reset the radio, press ⌊Menu⌋⌊ 6 ⌋⌊ 1 ⌋⌊Menu⌋ (plus version) or ⌊Menu⌋⌊ 5 ⌋ ⌊ 5 ⌋⌊Menu⌋ (non-plus version). Press ⌊Down⌋ to choose ALL. Press ⌊Menu⌋. You will see SURE? in the display. (If the device has a password set, you will see - - - - - - and you must enter the correct 6-digit password to continue.) Push ⌊Menu⌋ to reset.

VFO reset

To reset the radio, press Menu 6 1 Menu (plus version) or Menu 5 5 Menu (non-plus version). Press Down to choose VFO if it isn't already selected. Press Menu. You will see SURE? in the display. (If the device has a password set, you will see - - - - - - and you must enter the correct 6-digit password to continue.) Push Menu to reset.

⚠ **WARNING:** In at least some configurations, the Wouxun KG-UV9D may permit you to transmit on business or public safety frequencies. Make sure you are in-band when transmitting.

Yaesu FT-11R

Radio Layout

Specs

Receivers Single receiver with priority channel option
Receives 110–138 MHz AM, 136–180 MHz FM
Transmits 144–148 MHz @ 5 W FM
Antenna connector BNC F on radio; needs BNC M antenna
Modes FM
Memory Channels 146 (71 if using alpha tags)
Power No DC input on radio; 4–12 V DC per spec
Model year 1993

Standard Tasks

Program frequency in the field

1. Press VFO to go to VFO mode if you aren't already there.

2. Enter frequency omitting first two digits (for example, enter 7240 for 147.240).

3. Press F/M 1 to turn tone on (T in upper part of screen) or off. Press F/M again to leave function mode. If tone decode option is installed, radio will cycle through T (send tone on transmit), T SQ (send tone on transmit, require tone on receive to break squelch) and none.

4. Press F/M 6 to change repeater shift (+/-/none). Press F/M again to leave function mode.

5. Press F/M 2 to set tone frequency. Use knob or Up / Down to set tone desired CTCSS tone. Press 2 when done.

6. Press F/M 3 to set power level (as described below, "Change power in the field.")

7. Hold F/M for at least half a second to start write.

8. Scroll to desired memory (using knob or Up / Down).

9. Press and release F/M to write.

10. Press MR to go to memory mode.

11. For odd splits, program the receive frequency as described above. Then press VFO to go back to VFO mode, enter transmit frequency on keypad omitting first two digits (7840 for 147.840). Hold F/M for half a second. Scroll to memory that holds RX frequency using knob or Up / Down. Hold PTT (does not transmit) and press F/M to write transmit frequency. Successfully written memory will show +- to indicate odd split.

Lock/unlock radio

Flip lock switch up to lock (screen shows KL for keyboard lock, VL for volume lock, DL for knob lock and PL for PTT lock). Flip down to unlock.

Check repeater input frequency

Press [F/M] [9] to switch between reverse and regular modes.

Change power in the field

1. Press and release [F/M] [3]. You'll see either LOW (0.3–3 W) or HIGH (1.5–5 W), depending on battery installed). Press [3] to change between them.

2. On LOW, press [Up] / [Down] to choose between LOW 1 (0.3 W), LOW 2 (1.5 W) and LOW 3 (1.5 or 3 W, depending on battery installed).

3. Wait three seconds or press and release [PTT] to set (does not transmit).

Adjust volume

Press [Volume Up] or [Volume Down] to adjust volume.

Adjust squelch

Press [F/M]. Within three seconds, press [Volume Up] or [Volume Down] to adjust squelch.

Weird Modes

If radio is blinking LOW

Battery overheat temperature sensor has triggered. Radio automatically switches to low power when that happens. Transmit less and wait for the battery to cool down.

If radio doesn't receive when someone else transmits

This radio has a DTMF squelch option. Press [PAGE] repeatedly until you don't see PAGE, T.PAGE or CODE in the display.

If knob changes volume and not frequency

Press and hold both [Vol Up] and [Vol Down] when powering on.

If entering frequency as 4 digits doesn't work

Radio may be in extended frequency range mode. (See below.) Enter as 6 digits: 147200 for 147.200.

If radio displays ERR when PTT pressed

The radio displays ERR when you try to transmit out of band. Check frequency, repeater shift and repeater offset.

Useful Information

The radio can transmit CTCSS tones as shipped, but must have the FTS-26 option installed to do tone decoding / tone squelch.

Default offset can be changed. Press [F/M] [0] to enter set mode. Use knob to scroll to menu item 6 (usually 0.600). Use [Up] / [Down] to change. Press [PTT] (does not transmit) to save.

Lock type can be changed. Press [F/M] [0] to enter set mode. Use knob to scroll to menu item 5 (LOCK). Use [Up] to change PTT lock (PL) and keyboard lock (KL). Use [Down] to change volume lock (VL) and knob lock (DL). Press [PTT] (does not transmit) to save.

Extended frequency range mode: to set RX frequencies to extended range (110–180 MHz), press and hold both [Up] and [Down] while turning radio on. Frequencies below 136 MHz will be AM.

Factory reset

To reset the radio, hold down [MR], [VFO] and [2] while turning the radio on. This will reset all settings and memories.

Yaesu FT-209R/FT-209RH

Radio Layout

RF Power   Squelch

Power/Volume

PTT

M

MR

C

Rev

D

F

Lock

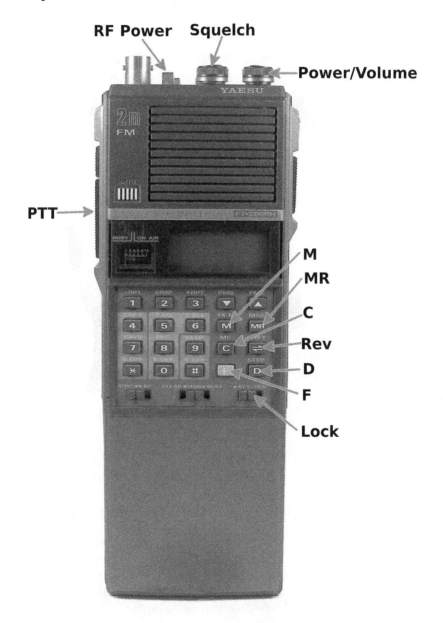

Specs

Receivers Single receiver
Receives 144–148 MHz FM
Transmits 144–148 MHz @ 3.5 W FM (209R), 144–148 MHz @ 5 W FM
(209RH)
Antenna connector BNC F on radio; needs BNC M antenna
Modes FM
Memory Channels 10
Power 6–15 V DC in spec, nominally 12 V DC on radio. Power adapter on
FNB-4 battery: IEC 60130-10 barrel style type A, 5.5mm OD, 2.5mm
ID plug, **center negative**. Charging plug on FNB-4 battery: 2.5mm
tip/sleeve phone plug, tip positive.
Model year 1985

Standard Tasks

Program frequency in the field

1. Decide which memory you want to write to (0–9). Note that 0 is special (can't be deleted, and is used for calling channel). In these instructions, $\boxed{n}$ represents the number key on the keyboard which
   corresponds to that memory.

2. Enter last four digits of frequency (4390 for 144.390). On B and C
   versions, the last digit is assumed (enter 438 for 144.3875). Then
   press $\boxed{D}$ to set the frequency. Press $\boxed{n}$ and then $\boxed{M}$ to write the
   frequency.

3. To set repeater shift, press $\boxed{F}$ then press $\boxed{1}$ (-RPT) or $\boxed{3}$ (+RPT)
   for negative or positive shift. Then press $\boxed{n}$ and then $\boxed{M}$ to write
   the shift.

4. To set non-standard shift, skip step 3 above and instead enter the
   transmit frequency you want. Then press $\boxed{D}$, $\boxed{n}$ $\boxed{F}$ and then $\boxed{M}$
   (TXM).

5. To select tone frequency, pick the correct one or two digit value from
   the table below, enter it, then press $\boxed{F}$ and then $\boxed{6}$ (T SET). Then
   press $\boxed{n}$ and then $\boxed{M}$ to write that to memory.

6. To activate tone encode, press $\boxed{F}$ and then $\boxed{8}$ (ENC). (For tone
   squelch, press $\boxed{F}$ and then $\boxed{5}$ (T SQ) instead.) Then press $\boxed{n}$ and
   then $\boxed{M}$ to write that to memory.

7. Press $\boxed{n}$ and then $\boxed{MR}$ to recall it.

Lock/unlock radio

Slide the keylock switch on the radio front panel to the right to lock. Slide
to the left to unlock.

Check repeater input frequency

Press and release Rev to switch between reverse and regular modes

Change power in the field

There is an RF pushbutton switch on top of the radio. Press it down low power (350 mW on FT-209R, 500 mW on FT-209RH). Press it up for high power (3.5 W on FT-209R, 5 W on FT-209RH).

Adjust volume

Rotate the right power/volume knob to adjust volume.

Adjust squelch

Rotate the left squelch knob to adjust RF squelch.

Weird Modes

Can't enter some frequencies

The radio comes in four versions (A, B, C and E). The A and E versions have 5 kHz channel spacing. The B and C versions have 12.5 kHz channel spacing, and will not let you enter any frequency which has a 4 or 9 in the 10 kHz digit, since those don't correspond to 12.5 kHz channel spacing. There is no way to enter those frequencies on the B and C versions.

Radio transmits unexpectedly

If you're using a headset, be sure you don't have VOX turned on. Press the VOX ON switch until it is up (off). VOX will not work without a headset.

1000 Hz tone when transmitting

In addition to the regular CTCSS tones, the FT-209 includes a test tone of 1000 Hz. Switch to a different tone (or none at all) to get rid of it.

Useful Information

The B version has a frequency range of 144–146 MHz rather than 144–148 MHz. The radio requires a FTS-6 tone board to generate CTCSS tones. Only the A version of the radio supports this board. (The A version can be recognized because it does not have a Burst above the PTT.)

DTMF is supported only on the A version of the receiver.

There is a switch on the radio underneath the battery pack that doubles channel spacing (goes from 5 to 10 kHz on A and E, 12.5 to 25 on B and C). You must remove the battery pack to see this switch, and will need a paperclip or other small object to press it.

The switches on the front panel are: S/PO: meter shows signal strength on receive, power on transmit. BC: meter shows battery. Clear: scan stops on empty channel. Man: scan stops when scan key is released. Busy: scan stops on busy channel.

To delete any memory except 0: press $\boxed{n}$ then press $\boxed{MR}$, then press $\boxed{F}$, $\boxed{C}$, $\boxed{D}$.

To see which memories are in use, press $\boxed{D}$, $\boxed{M}$, $\boxed{M}$.

CTCSS tones (Hz) and their FT-209 code values

67.0: 34	91.5: 42	123.0: 13	162.2: 21	218.1: 29
71.9: 35	94.8: 6	127.3: 14	167.9: 22	225.7: 30
74.4: 36	100.0: 7	131.8: 15	173.8: 23	233.6: 31
77.0: 37	103.5: 8	136.5: 16	179.9: 24	241.8: 32
79.7: 38	107.2: 9	141.3: 17	186.2: 25	250.3: 55
82.5: 39	110.9: 10	146.2: 18	192.8: 26	
85.4: 40	114.8: 11	151.4: 19	203.5: 27	
88.5: 41	118.8: 121	156.7: 20	210.7: 28	1000: 63

Factory reset

To reset, insert a paperclip in the hole on the back of the case. This resets all memories and settings.

Yaesu FT-25

Radio Layout

Power/Volume

Emergency

PTT

Moni

F

P2
P1
P3
P4

Up

Down

*

#

Specs

Receivers Single receiver with priority channel option
Receives 65–108 MHz WFM, 136–174 MHz FM
Transmits 144–148 MHz @ 5 W FM
Antenna connector SMA **M** on radio; needs SMA **F** antenna
Modes FM
Memory Channels 200
Power No power input on radio, nominally 7.4 V DC
Model year 2017

Standard Tasks

Program frequency in the field

1. Enter VFO mode by pressing [#] if you're not already there. (Display shows M if you're in memory mode.)

2. Enter the frequency on the keypad (144390 for 144.390 MHz).

3. Hold [F] to enter menu mode.

4. To set the tone, use [Up] and [Down] to scroll to item 8, CTCSS. Press [F] to select TX if desired, then [Up] / [Down] to adjust. Press [F] to set. Similarly, use [Up] / [Down] to select RX if desired, then [Up] / [Down] to adjust. Press [F] to set.

5. To set shift, use [Up] and [Down] to scroll to item 24, Repeater. Press [F] to select Mode. Use [Up] / [Down] to switch between Simplex, +REP (positive offset) and -REP (negative offset). Press [F] to set.

6. If you need to adjust repeater offset frequency, use [Down] to select Shift. Press [F] to change, and use [Up] / [Down] to adjust. Press [F] to set when done.

7. To set tone type, use [Up] / [Down] to scroll to item 29, SQL Type. Press [F] to adjust. Use [Up] / [Down] to select from R-TONE (receive squelch), T-TONE (transmit tone), TSQL (transmit tone and receive tone squelch), REV TN (squelch if tone received), DCS (digital coded squelch), PAGER (unsquelch if two CTCSS tones received) , and OFF (no tone). Press [F] to set.

8. To adjust power, use [Up] / [Down] to go to item 32, TX PWR. Press [F] to adjust. Use [Up] / [Down] to choose from HI(5 W), MID(2.5 W) or LOW(0.5 W). Press [F] to set.

9. Hold [F] to exit the menu.

10. Hold [*] to start writing to memory.

11. Use [Up] / [Down] to select the memory.

12. hold [*] to complete writing to memory.

13. Press [*] to switch to memory mode if you're not already there.

14. Use [Up] / [Down] to scroll to the desired memory.

Lock/unlock radio

Hold [6] to lock or unlock.

Check repeater input frequency

Press [F] [P4] to enter or exit reverse mode. Shift will blink when you're in reverse mode.

Change power in the field

To adjust power, hold [F] to enter the menu. Use [Up] / [Down] to go to item 32, TX PWR. Press [F] to adjust. Use [Up] / [Down] to choose from HI(5 W), MID(2.5 W) or LOW(0.5 W). Press [F] to set. Hold [F] to exit the menu.

Adjust volume

Rotate the knob to adjust volume.

Adjust squelch

Hold [F] to enter hte menu. Scroll to item 26, RF SQL. Press [F] to adjust. Use [Up] / [Down] to go from OFF, S-1 through S-8, or S-FULL. Press [F] to set. Hold [F] to exit the menu.

Weird Modes

Radio is beeping continually

You are in emergency mode. Turn the radio off and then on. If you hold [Emergency] for three seconds, you will enter emergency mode. This causes a continual beep and wil transmit on the home frequency.

Can't enter VFO mode

This radio has a memory-only mode. To exit, turn the radio off. Then hold down [Moni] and [PTT] (does not transmit) while turning the radio on. Use [Down] to select F5:MEM-ONLY. Press [F] to toggle.

Can't enter VHF frequencies

This radio has an option to use FM only. To adjust it, turn the radio off. Then hold down [Moni] and [PTT] (does not transmit) while turning the radio on. Use [Down] to select F7:FM-ONLY. Press [F] to toggle.

Radio shows OUT.RNG or IN.RNG

This means you have ARTS mode enabled. Hold [2] to exit.

Radio shows - - - - on power on

The password has been enabled. Enter the password (four digits) on the keypad. If you have forgotten the password, you can disable with a full reset (see below).

Useful Information

This radio has four programmable buttons which remember the radio state. Rather than programming a memory, once you've set up the frequency, hold [P1], [P2], [P3] or [P4] to program the button. Then later you can press [P1], [P2], [P3] or [P4] to recall the state (including frequency / tone info).

Factory reset

To reset, hold down [Moni] and [PTT] (does not transmit) while turning the radio on. Scroll to F4 ALL RESET (full reset—reset everything). Press [F] to reset.

Memory reset

To reset, hold down [Moni] and [PTT] (does not transmit) while turning the radio on. Scroll to F2 MEM RESET (reset memories). Press [F] to reset.

Memory bank reset

To reset, hold down [Moni] and [PTT] (does not transmit) while turning the radio on. Scroll to F3 BANK RESET (reset memory banks). Press [F] to reset.

Settings reset

To reset, hold down [Moni] and [PTT] (does not transmit) while turning the radio on. Scroll to F1 SET RESET (reset settings). Press [F] to reset.

Yaesu FT-252

Radio Layout

Knob

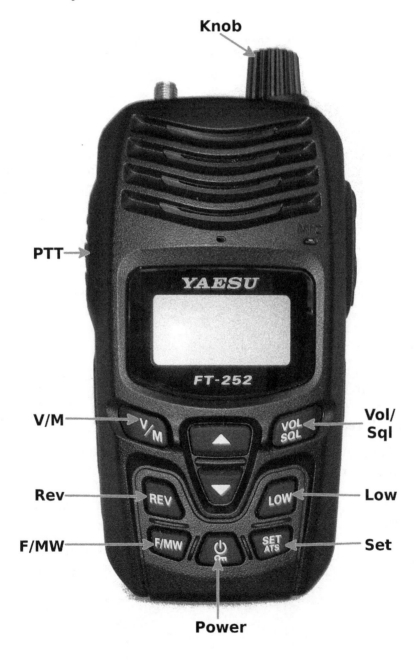

PTT

V/M

Rev

F/MW

Vol/Sql

Low

Set

Power

Specs

Receivers Single receiver
Receives 136–174 MHz FM
Transmits 144–148 MHz @ 5 W FM
Antenna connector SMA F on radio; needs SMA M antenna
Modes FM
Memory Channels 200
Power Nominally 5–10 V DC, but charger is 10.5 V DC, EIAJ-02 barrel
    style, 4mm OD, 1.7mm ID plug, center positive
Model year 2013

Standard Tasks

Program frequency in the field

1. Press ⟨V/M⟩ to go to VFO mode if you aren't already there.

2. Use knob to select frequency. Press ⟨F/MW⟩ to adjust in 1 MHz steps.

3. Press ⟨Set⟩ to enter menu.

4. Scroll using the knob to Menu `42:TN FRQ`. Press ⟨Set⟩ to adjust tone frequency. Use the scroll knob for adjustment, and press ⟨Set⟩ when done. (Use `13:DCS.COD` for DCS tones.)

5. Scroll to Menu `40:SQL.TYP` to adjust squelch type. Press ⟨Set⟩ to adjust (`TONE` (CTCSS tone on transmit), `TSQL` (tone on transmit and receive), `REV TN` (squelch on CTCSS tone receive), `DCS`, `D` (DCS encode), `T DCS` (CTCSS transmit, DCS for receive), `D TSQL` (DCS transmit, CTCSS receive) or `OFF`). Use the scroll knob for adjustment, and press ⟨Set⟩ when done.

6. Scroll to Menu `32:RPT.MOD` to adjust repeater shift (`RPT. +`, `RPT.OFF`, or `RPT. -`). Press ⟨Set⟩ to adjust. Use the scroll knob for adjustment, and press ⟨Set⟩ when done.

7. If necessary, scroll to Menu `28:R SHFT` to adjust repeater shift. Press ⟨Set⟩ then use knob to adjust repeater shift. Press ⟨Set⟩ when done.

8. Press ⟨F/MW⟩ to leave menu mode.

9. Press ⟨Low⟩ repeatedly to select power level. Options are (`HIGH` 5 W/`MID` 2 W/`LOW` 0.5 W). Wait about two seconds for the mode to be set.

10. Hold ⟨F/MW⟩ for at least half a second to start write.

11. Scroll to desired memory.

12. Press and release ⟨F/MW⟩ to write.

Lock/unlock radio

Press [Power] to lock or unlock.

Check repeater input frequency

Press and release [Rev] to switch between reverse and regular modes. For this to work, the radio must have Menu 30:REV/HM set to REV mode.

Change power in the field

Press [Low] repeatedly to select power level. Options are (HIGH 5 W/MID 2 W/LOW 0.5 W). Wait about two seconds for the mode to be set.

Adjust volume

Press [Vol/Sql] until you see VOL, then rotate knob to adjust volume. The radio automatically leaves this mode after about three seconds with no action.

Adjust squelch

Press [Vol/Sql] until you see SQL, then rotate knob to adjust volume. The radio automatically leaves this mode after about three seconds with no action.

Weird Modes

Radio is beeping continually

You are in emergency mode. Turn the radio off and then on.

Radio shows OUT.RNG, IN.RNG or SYNC

This means you have ATS mode enabled. Press [F/MW] and then [Set] to disable.

Radio shows - - - - on power on

The password has been enabled. Use the knob to select and press [F/MW] to enter each character at a time. If you have forgotten the password, you can disable with a full reset (see below).

Can't get to VFO mode

The radio has a memory-only mode. To disable, hold [V/M] while turning the radio on, then select F5:M-ONLY, then press [Set]. This toggles from memory-only mode to regular mode.

Useful Information

Hold Power for about two seconds to turn off or on.

If you hold Set for one second, you will enter emergency mode. This causes a continual beep and can transmit if configured to do so using Menu 19:EMG S. Options are EMG.BEP (beep), EMG.LMP (flash the display), EMG.B+L (beep + display), EMG.CWT (transmit SOS), EMG.C+B (transmit + beep), EMG.C+L (transmit + display), EMG.ALL (transmit + beep + display) and (mercifully) OFF.

In order to see some transmit tone options, Menu 39:SPLIT must be set to SPL ON.

This radio has no way to program it other than to do so by the front panel. Be cautious about resetting memories, since the radio user will have to manually re-enter everything.

Factory reset

To reset, hold down V/M and turn the radio on. Scroll to F4 ALLRST (full reset—reset everything). Press Set to reset.

Memory reset

To reset, hold down V/M and turn the radio on. Scroll to F2 MEMRST (reset memories). Press Set to reset.

Memory bank reset

To reset, hold down V/M and turn the radio on. Scroll to F3 MB RST (reset memory banks). Press Set to reset.

Settings reset

To reset, hold down V/M and turn the radio on. Scroll to F1 SETRST (reset settings). Press Set to reset.

Yaesu FT-257

Radio Layout

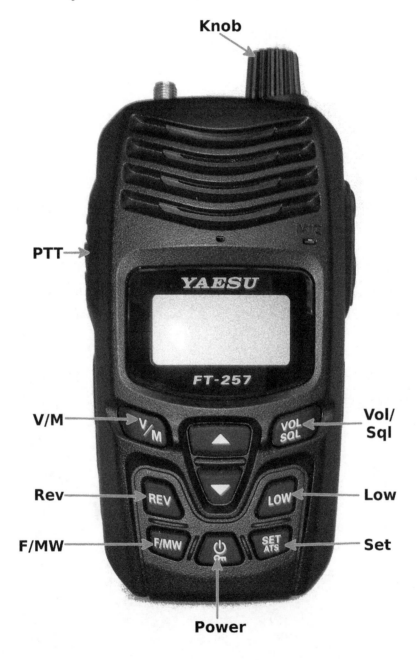

Knob

PTT

V/M

Rev

F/MW

Vol/ Sql

Low

Set

Power

Specs

Receivers Single receiver
Receives 400–480 MHz FM
Transmits 430–450 MHz @ 5 W FM
Antenna connector SMA F on radio; needs SMA M antenna
Modes FM
Memory Channels 200
Power Nominally 5–10 V DC, but charger is 10.5 V DC, EIAJ-02 barrel
   style, 4mm OD, 1.7mm ID plug, center positive
Model year 2013

Standard Tasks

Program frequency in the field

1. Press [V/M] to go to VFO mode if you aren't already there.

2. Use knob to select frequency. Press [F/MW] to adjust in 1 MHz steps.

3. Press [Set] to enter menu.

4. Scroll using the knob to Menu 42:TN FRQ. Press [Set] to adjust tone frequency. Use the scroll knob for adjustment, and press [Set] when done. (Use 13:DCS.COD for DCS tones.)

5. Scroll to Menu 40:SQL.TYP to adjust squelch type. Press [Set] to adjust (TONE (CTCSS tone on transmit), TSQL (tone on transmit and receive), REV TN (squelch on CTCSS tone receive), DCS, D (DCS encode), T DCS (CTCSS transmit, DCS for receive), D TSQL (DCS transmit, CTCSS receive) or OFF). Use the scroll knob for adjustment, and press [Set] when done.

6. Scroll to Menu 32:RPT.MOD to adjust repeater shift (RPT. +, RPT.OFF, or RPT. -). Press [Set] to adjust. Use the scroll knob for adjustment, and press [Set] when done.

7. If necessary, scroll to Menu 28:R SHFT to adjust repeater shift. Press [Set] then use knob to adjust repeater shift. Press [Set] when done.

8. Press [F/MW] to leave menu mode.

9. Press [Low] repeatedly to select power level. Options are (HIGH 5 W/MID 2 W/LOW 0.5 W). Wait about two seconds for the mode to be set.

10. Hold [F/MW] for at least half a second to start write.

11. Scroll to desired memory.

12. Press and release [F/MW] to write.

Lock/unlock radio

Press [Power] to lock or unlock.

Check repeater input frequency

Press and release [Rev] to switch between reverse and regular modes. For this to work, the radio must have Menu 30:REV/HM set to REV mode.

Change power in the field

Press [Low] repeatedly to select power level. Options are (HIGH 5 W/MID 2 W/LOW 0.5 W). Wait about two seconds for the mode to be set.

Adjust volume

Press [Vol/Sql] until you see VOL, then rotate knob to adjust volume. The radio automatically leaves this mode after about three seconds with no action.

Adjust squelch

Press [Vol/Sql] until you see SQL, then rotate knob to adjust volume. The radio automatically leaves this mode after about three seconds with no action.

Weird Modes

Radio is beeping continually

You are in emergency mode. Turn the radio off and then on.

Radio shows OUT.RNG, IN.RNG or SYNC

This means you have ATS mode enabled. Press [F/MW] and then [Set] to disable.

Radio shows - - - - on power on

The password has been enabled. Use the knob to select and press [F/MW] to enter each character at a time. If you have forgotten the password, you can disable with a full reset (see below).

Can't get to VFO mode

The radio has a memory-only mode. To disable, hold [V/M] while turning the radio on, then select F5:M-ONLY, then press [Set]. This toggles from memory-only mode to regular mode.

Useful Information

Hold [Power] for about two seconds to turn off or on.

If you hold [Set] for one second, you will enter emergency mode. This causes a continual beep and can transmit if configured to do so using Menu 19:EMG S. Options are EMG.BEP (beep), EMG.LMP (flash the display), EMG.B+L (beep + display), EMG.CWT (transmit SOS), EMG.C+B (transmit + beep), EMG.C+L (transmit + display), EMG.ALL (transmit + beep + display) and (mercifully) OFF.

In order to see some transmit tone options, Menu 39:SPLIT must be set to SPL ON.

This radio has no way to program it other than to do so by the front panel. Be cautious about resetting memories, since the radio user will have to manually re-enter everything.

Factory reset

To reset, hold down [V/M] and turn the radio on. Scroll to F4 ALLRST (full reset—reset everything). Press [Set] to reset.

Memory reset

To reset, hold down [V/M] and turn the radio on. Scroll to F2 MEMRST (reset memories). Press [Set] to reset.

Memory bank reset

To reset, hold down [V/M] and turn the radio on. Scroll to F3 MB RST (reset memory banks). Press [Set] to reset.

Settings reset

To reset, hold down [V/M] and turn the radio on. Scroll to F1 SETRST (reset settings). Press [Set] to reset.

Yaesu FT-270

Radio Layout

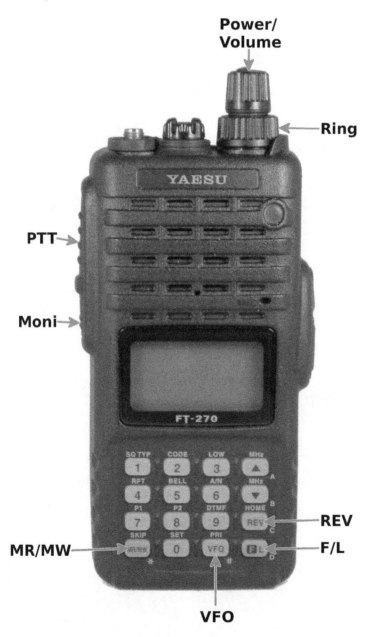

Power/
Volume

Ring

PTT

Moni

MR/MW

REV

F/L

VFO

Specs

Receivers Single receiver, option of priority channel or dual watch (first to break squelch wins)
Receives 136–174 MHz FM
Transmits 144–148 MHz @ 5 W FM
Antenna connector SMA F on radio; needs SMA M antenna
Modes FM
Memory Channels 200
Power 6.0–16.0 V DC, EIAJ-02 barrel style, 4mm OD, 1.7mm ID plug, center positive
Model year 2009

Standard Tasks

Program frequency in the field

1. Press ⟨VFO⟩ to go to VFO mode if you aren't already there. (Display will show -A- or -b- in VFO mode, and a memory number in memory mode.)

2. Enter frequency (144390 for 144.390 MHz).

3. To set tone, press ⟨F/L⟩ ⟨1⟩ (SQ TYPE). Use the ring to adjust. Values are OFF (no tone), TONE (tone on transmit), TSQL (tone sent on transmit and required for receive), REV TN (tone causes radio not to receive), DCS (digital coded squelch), ECS (enhanced paging squelch). Press ⟨PTT⟩ (does not transmit) to set.

4. To set offset, press ⟨F/L⟩ ⟨4⟩ (RPT). Use the ring to adjust. Values are RPT. + (positive offset), RPT.OFF (no offset) and RPT. – (negative offset). Press ⟨F/L⟩ to set.

5. To set tone frequency, press ⟨F/L⟩ ⟨0⟩ to enter the set menu. Use the ring to scroll to item 46, TN FRQ. Press ⟨F/L⟩ to change. Use the ring to adjust, and press ⟨F/L⟩ to set.

6. To set offset frequency, press ⟨F/L⟩ ⟨0⟩ to enter the set menu. Use the ring to scroll to item 41, SHIFT. Press ⟨F/L⟩ to change. Use the ring to adjust, and press ⟨F/L⟩ to set.

7. Press ⟨PTT⟩ (does not transmit) to leave menu mode.

8. To set transmit power, press ⟨F/L⟩ ⟨3⟩ (LOW). Use the ring to select the level you want: (HIGH (5 W), MID (2 W), or LOW (0.5 W). Press and release ⟨F/L⟩ to set the power level.

9. Hold ⟨MR/MW⟩ for at least one second to start write.

10. Use the ring to select the desired memory to write.

11. Press ⌊MR/MW⌋ to write.

12. Press ⌊MR/MW⌋ to go to memory mode; scroll to the memory you just wrote.

Lock/unlock radio

Hold ⌊F/L⌋ for one second to lock or unlock.

Check repeater input frequency

Press and release ⌊REV⌋ to switch transmit and receive frequencies. (The repeater offset will blink when you are reversed.)

Change power in the field

To set transmit power, press ⌊F/L⌋ ⌊3⌋ (LOW). Use the ring to select the level you want: (HIGH (5 W), MID (2 W), or LOW (0.5 W). Press and release ⌊F/L⌋ to set the power level.

Adjust volume

Rotate the center power/volume knob to adjust volume.

Adjust squelch

Press ⌊F/L⌋ ⌊Moni⌋. Rotate the outer ring to adjust squelch (1–15). Press ⌊PTT⌋ (does not transmit) to save.

Weird Modes

Radio shows OUT.RNG

This means you have ARTS mode enabled. Press ⌊F/L⌋ to exit ARTS mode.

Can't enter VFO mode

This radio has a memory-only mode. To enable/disable it, hold down ⌊Moni⌋ while turning the radio on. Scroll to F5 M-ONLY and press ⌊F/L⌋.

Radio doesn't hear transmissions

This radio has a feature to prevent signals from opening squelch unless they have a certain S-meter reading. To change this, press ⌊F/L⌋ ⌊0⌋. Rotate the outer ring to choose menu 34, RF SQL. Press ⌊F/L⌋ to adjust. Rotate the outer ring to set to OFF. Press ⌊PTT⌋ (does not transmit) to save.

Useful Information

Factory reset

To reset the radio, hold down [Moni] while turning the radio on. Use the outer ring to select F4 ALLRST (reset everything). Press [F/L] to reset.

Memory reset

To reset the radio, hold down [Moni] while turning the radio on. Use the outer ring to select F2 MEMRST (clear memories). Press [F/L] to reset.

Memory bank reset

To reset the radio, hold down [Moni] while turning the radio on. Use the outer ring to select F3 MB RST (clear memory banks). Press [F/L] to reset.

Settings reset

To reset the radio, hold down [Moni] while turning the radio on. Use the outer ring to select F1 SETRST (reset settings). Press [F/L] to reset.

Yaesu FT-2D

Radio Layout

Specs

Receivers Single receiver, dual watch (first to break squelch wins)

Receives 522–1720 kHz AM / FM, 1.8–774 MHz AM / FM, 803–999 MHz AM / FM

Transmits 144–148 MHz @ 5 W FM / C4FM, 430–450 MHz @ 5 W FM / C4FM

Antenna connector SMA F on radio; needs SMA M antenna

Modes FM, C4FM Fusion

Memory Channels 900

Power 4–14 V DC, EIAJ-02 barrel style, 4mm OD, 1.7mm ID plug, center positive

Model year 2015

Standard Tasks

Program frequency in the field

1. Press [V/M] to go to VFO mode if you aren't already there. The dipslay will show VFO.

2. Press [Band] repeatedly to cycle through bands and select the appropriate one.

3. Press [Disp] until you see the soft button [F MW] on the screen.

4. To set the frequency, tap the frequency on the screen. The radio will then display a soft keyboard which lets you enter frequency. Enter 144390 for 144.390 MHz.

5. Press the soft button [MODE] until you see FM (other options are DN (Fusion digital), VW (full bandwidth digital voice), DW (full bandwidth digital data). If you see a mode with a bar over it, that is an automatically selected mode.

6. To set the tone, press the soft button [F MW] and then press the soft button [SQ TYPE]. Use the knob to cycle through OFF (no tone), TONE (CTCSS tone on transmit), TONE SQL (CTCSS tone on receive), DCS (digital coded squelch), REV TONE (squelch if tone received), PR FREQ ("no-communication squelch" with a frequency) and PAGER (detect two CTCSS tones). Press [PTT] (does not transmit) to set.

7. To set repeater shift, hold [Disp] for at least one second to enter set mode. Press the soft button [CONFIG]. Use the knob to scroll to 14RPT ARS and touch that to turn it on or off. Use the knob to scroll to 15RPT SHIFT and touch that to cycle through -RPT (negative shift), +RPT (positive shift) and SIMPLEX (no shift). Press [PTT] (does not transmit) to set.

8. If you need to change the repeater offset frequency, hold [Disp] for at least one second to enter set mode. Press the soft button [CONFIG].

Use the knob to scroll to 16RPT SHIFT FREQ, touch that, and use the knob to adjust. Press PTT (does not transmit) to set.

9. To set the tone frequency, hold Disp for at least one second to enter set mode. Press the soft button SIGNALING. Use the knob to scroll to 12TONE SQL FREQ and touch that to change. Use the knob to adjust the value. Press PTT (does not transmit) to set.

10. To set transmit power, press the soft F MW and then the soft button TXPWR. Use the knob to select from HIGH, LOW1, LOW2, LOW3. Use PTT (does not transmit) to set.

11. Hold the soft button F MW for about one second, but release it quickly after that.

12. Use the knob to select the memory to write.

13. Press the soft button M.WRITE to save to memory.

14. Press V/M to go to memory mode.

15. Use the knob to select the memory.

Lock/unlock radio

Press Power to lock or unlock.

Check repeater input frequency

Press the soft button F MW and then the soft button REV to switch to reverse input. (In reverse mode the offset indicator will blink.) Repeat to switch back.

Change power in the field

To set transmit power, press the soft F MW and then the soft button TXPWR. Use the knob to select from HIGH (5 W), LOW1 (0.1 W), LOW2 (1 W), and LOW3 (2.5 W). Use PTT (does not transmit) to set.

Adjust volume

Rotate the ring to adjust the volume.

Adjust squelch

Press SQL. Use the knob to adjust squelch from 0–15. Press SQL to set.

Weird Modes

Radio doesn't transmit

This radio has a PTT lock. Unlock the radio to transmit.

Signals are weak

This radio has a built-in signal attenuator. To enable or disable, hold [Disp] to enter set mode, then push the soft button [TX/RX]. Press the soft button [MODE] and then the soft button [ANTENNA ATT]. Use the knob to turn it on/off, and then press [PTT] (does not transmit) to set.

Radio shows GM and some functions don't work

You have activated the Group Monitor mode. Press [GM] to turn it off.

Radio shows WIRES-X and some functions don't work

You have activated the WIRES-X mode. Hold [X] to turn it off. Press [X] to turn it on.

Useful Information

Hold [Power] for about two seconds to turn off or on.

Factory reset

To reset everything, turn the radio off. Then hold [Band], [Back] and [Disp] while turning the radio on. You will see ALL RESET?. Press the soft button [OK] to reset.

Settings reset

To reset most menu settings (but keep programmed memories), turn the radio off. Then hold [Back] and [Disp] while turning the radio on. You will see SET MODE RESET?. Press the soft button [OK] to reset.

Yaesu FT-50

Radio Layout

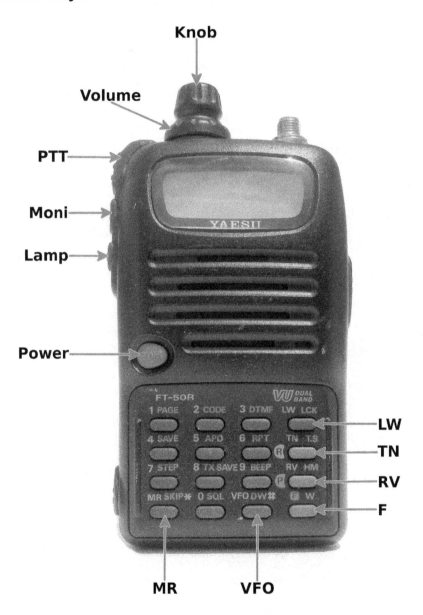

Specs

Receivers Single receiver with priority channel option (memory 1 in memory mode)
Receives 76–200 MHz AM/FM, 300–400 MHz AM / FM, 400–540 MHz AM / FM, 590–999 MHz AM / FM
Transmits 144–148 MHz @ 5 W FM, 430–450 MHZ @ 5 W FM
Antenna connector SMA F on radio; needs SMA M antenna
Modes FM
Memory Channels 99
Power 5–13 V DC, EIAJ-02 barrel style, 4mm OD, 1.7mm ID plug, center positive
Model year 1996

Standard Tasks

Program frequency in the field

1. Press ⎡VFO⎤ to go to VFO mode if you aren't already there. There are two VFO modes (A and B); you can be in either.

2. Make sure your step size is set to 5. Press ⎡F⎤ then ⎡7⎤ to adjust step size, then rotate knob to select STEP 5.0. Press ⎡7⎤ to exit.

3. Enter frequency (144390 for 144.390 MHz). (If step size is greater than 5, you might not have to / be able to enter the last digit.)

4. The radio has automatic repeater shift (ARS). To override that, hold ⎡Knob⎤ for half a second, scroll to RPTR -06- and press ⎡Knob⎤. Rotate to select from -RPT (negative shift), SIMP (simplex) and +RPT (positive shift). Press ⎡PTT⎤ (does not transmit) to leave menu and save shift.

5. To set tone type, press ⎡TN⎤ to cycle through T (tone), T SQ (tone squelch—only if FTT-12 is installed) and DCS (DCS). Press ⎡F⎤ then ⎡TN⎤ to enter tone adjust mode. Scroll using the knob to the correct tone, then press ⎡TN⎤.

6. To set power level, hold ⎡Knob⎤ for half a second, then scroll to TXPO- -02-. Press ⎡Knob⎤ and scroll to select power level: HI 5 W / L1 0.1 W / L2 1 W / L3 2.8 W. Press ⎡PTT⎤ (does not transmit) to exit.

7. Hold ⎡F⎤ for at least half a second to start write.

8. Scroll to desired memory. Memories with data appear as CH-2; memories without data appear as CHu2.

9. Press and release ⎡F⎤ to write.

Lock/unlock radio

Press ⎡F⎤ then ⎡LW⎤ to lock or unlock. The type of lock/unlock can be changed with menu LOCK -17-. Lock types are keypad KL, knob DL (knob), PTT PL, and combinations of those.

Check repeater input frequency

Press and release RV to switch between reverse and regular modes.

Change power in the field

Hold Knob for half a second, then scroll to TXPO -02-. Press Knob and scroll to select power level: HI 5 W / L1 0.1 W / L2 1 W / L3 2.8 W. Press PTT (does not transmit) to exit.

Adjust volume

Rotate the outer volume ring to adjust volume.

Adjust squelch

Hold Knob for half a second. Scroll to SQL -01-. Press Knob and scroll to select squelch level (0–15). Press PTT (does not transmit) to exit.

Weird Modes

If someone says you have Wires mode turned on

This radio doesn't have Wires, but does have a DTMF paging mode. To exit it, press F and then repeatedly press 1 until you don't see CODE or PAGE on the bottom of the screen. Then press F.

Radio shows TRX, TX or RX

This means you have ARTS mode enabled. To disable, press F then hold Knob for half a second. Press F then press TN. Finally, press F repeatedly until you no longer see TRX, TX or RX. Then press TN.

Can't enter VFO mode

This radio has a memory-only mode. To exit it (or enter it again), turn the radio off. Then hold down PTT and Lamp and turn the radio on again.

Radio shows all LCD segments

This radio has an LCD test mode, entered by turning on the radio while holding Lamp. Turn the radio off and then back on to clear.

Automatic repeater shift is wrong

There is a modification to enable cell band receive for this radio. If that modification has been applied, ARS will be incorrect for North America (it will be European shifts) and offsets will be wrong. You will have to set the repeater shifts and offsets manually.

Useful Information

Some versions of this radio have the FTT-12 option installed. This enables brief recordings as well as tone squelch mode. Radios with this option installed have an "R" marking to the left of TN and a "P" marking to the left of RV on the keypad.

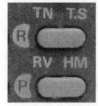

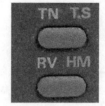

FTT-12 installed   FTT-12 not installed

To turn the radio on or off, hold Power for about one second.

The radio has an extended receive mode. To enter it, hold down Knob and Lamp while powering on. After receive has been extended, performing this procedure will reset everything.

The radio has a game mode. To enter it, hold down Knob and MR while powering on. Select speed with the knob, then press PTT (does not transmit) to start. A number will display and scroll to the right. Enter the number which will cause the total to be 10, then press F (use 1 0 for 0). Turn the radio off to exit.

This radio has a service mode which can be entered by holding Knob, PTT and Lamp while turning on the radio. Be cautious since it is possible to convert a working radio into a non-working radio in this mode. Turn the radio off to exit the mode.

Factory reset

To reset everything, hold down Moni and Knob then turn the radio on. Display will show ALRST PrS F. Press F to reset.

Yaesu FT-530

Radio Layout

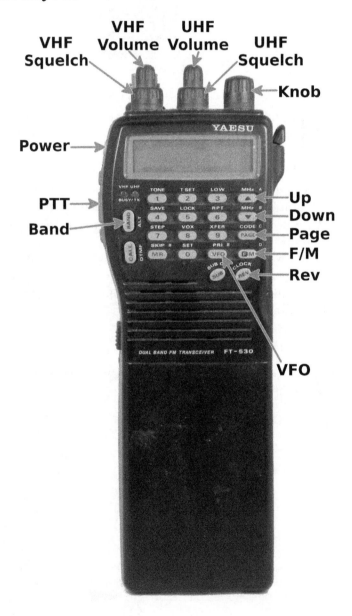

Specs

Receivers Two independent receivers, simultaneous receive
Receives 130–174 MHz AM/FM, 430–450 MHz FM
Transmits 144–148 MHz @ 5 W FM, 430–450 MHZ @ 5 W FM
Antenna connector BNC F on radio; needs BNC M antenna
Modes FM
Memory Channels 76 total—38 on left side, 38 on right side
Power 5–13 V DC, EIAJ-02 barrel style, 4mm OD, 1.7mm ID plug, center positive
Model year 1992

Standard Tasks

Program frequency in the field

1. Press [Band] to select either the left band (2m) or the right band (70cm).

2. Press [VFO] to go to VFO mode if you aren't already there. There are two VFO modes (A and B); you can be in either.

3. Enter frequency (144390 for 144.390 MHz).

4. To set offset frequency (for everything on that band), press [F/M] [0] [6]. Use the knob to set. Press [6] to save.

5. To set offset direction, press [F/M] then repeatedly press [6] to cycle through -, + and blank (no offset). Press [F/M] to set.

6. To set tone, press [F/M] then [2]. Use the knob to select tone. Press [2] to set.

7. To set tone type, first press [F/M] then repeatedly press [1] to cycle through T (tone), T SQ (tone squelch) and blank (no tone). Press [F/M] to set.

8. To set power level, press [F/M] then [3]. You will see either Hi or something that starts with L. Hi corresponds to high power (5 W on 12V DC, 2 W on 7.2V DC). If you are on L, you can use the knob to select from L1 (0.5 W), L2 (1.5 W) or L3 (2 W). Wait two seconds after setting level to go back to normal mode. The display will show LOW when you are using a low power setting, and nothing when you are using the high power setting.

9. Hold [F/M] for at least half a second to start write.

10. Scroll to desired memory. Memories with data are solid; memories without data are blinking.

11. Press and release [F/M] to write.

Lock/unlock radio

Press F/M then 5 repeatedly to lock PTT (PL), keyboard (KL) and knob (DL if enabled). Press F/M or wait two seconds to set.

Check repeater input frequency

Press and release Rev to switch between reverse and regular modes.

Change power in the field

To set power level, press F/M then 3. You will see either Hi or something that starts with L. Hi corresponds to high power (5 W on 12V DC, 2 W on 7.2V DC). If you are on L, you can use the knob to select from L1 (0.5 W), L2 (1.5 W) or L3 (2 W). Wait two seconds after setting level to go back to normal mode. The display will show LOW when you are using a low power setting, and nothing when you are using the high power setting.

Adjust volume

Rotate the center volume knob for the correct side of the radio to adjust volume.

Adjust squelch

Rotate the outer squelch knob for the correct side of the radio to adjust squelch.

Weird Modes

Radio shows blinking LOW

The radio will display a blinking LOW if it is overheating. Stop transmitting / move to a lower power level.

Received audio bad

The radio can be switched into AM mode for 2m. To enable/disable AM, press F/M 0 then repeatedly press F/M VFO until you see A3 on (AM on) or A3 OFF (AM off). Press 0 to set.

Radio plays annoying tones when you press any button

This can be disabled/enabled with F/M 2 F/M 2.

Radio plays tones when keyed up (PTT)

The radio has a number of page / code modes. Repeatedly press Page until you don't see a bell symbol, CODE or PAGE in the display.

Useful Information

This radio has crossband repeat capability. (This is full crossband; whatever is transmitted on one frequency will be sent out on the other. Turn the volume knobs down to avoid feedback.) To enable, press [6] while turning on the radio. Whatever frequencies / tones that were in memory for VHF and UHF will be used to crossband repeat.

The radio will transmit VHF only on the left display, and transmit UHF only on the right display. You can receive either frequency on either display.

This radio has a clock. The clock display can be turned on or off with [F/M] [Rev] [5] [Rev].

The radio has automatic repeater shift (ARS). It is disabled by default; to toggle it press [F/M] [0] [6] [F/M]. You will see A when ARS is turned on. Press [6] to exit.

The radio has an extended receive mode that can be enabled after a hardware modification. If the modification has been done, you can enter extended receive by holding down [Up] and [Down] while powering on. After receive has been extended, performing this procedure will reset everything. After entering extended receive, program 110.000 in the left side L memory, 180.000 in the left side U memory, 300.000 in the right side L memory and 500.00 in the right side U memory.

Factory reset

To reset everything, hold down [MR] and [VFO] then turn the radio on. This will reset everything (including channel memories) without any prompts.

Yaesu FT-60

Radio Layout

Specs

Receivers Single receiver, dual watch (first to break squelch wins)
Receives 108–520 MHz AM/FM, 700–999 MHz AM/FM
Transmits 144–148 MHz @ 5 W FM, 430–450 MHZ @ 5 W FM
Antenna connector SMA F on radio; needs SMA M antenna
Modes FM
Memory Channels 1000
Power 6–16 V DC, EIAJ-02 barrel style, 4mm OD, 1.7mm ID plug, center
positive
Model year 2004

Standard Tasks

Program frequency in the field

1. Press $\boxed{\text{V/M}}$ to go to VFO mode if you aren't already there.

2. Enter frequency (144390 for 144.390 MHz).

3. Press $\boxed{\text{F/W}}$ $\boxed{0}$ to enter menu.

4. Scroll using the right knob to Menu 50:TN FRQ. Press $\boxed{\text{F/W}}$ to adjust tone frequency. Use the scroll knob for adjustment, and press $\boxed{\text{F/W}}$ when done.

5. Scroll to Menu 48:SQL.TYP. to adjust squelch type. Press $\boxed{\text{F/W}}$ to adjust (TONE, TSQL, REV TN, DCS or OFF). Use the scroll knob for adjustment, and press $\boxed{\text{F/W}}$ when done.

6. Scroll to Menu 38:RPT.MOD to set repeater shift (RPT. +/RPT.OFF / RPT. -). Press $\boxed{\text{F/W}}$ to adjust. Use the scroll knob for adjustment, and press $\boxed{\text{F/W}}$ when done.

7. Hold $\boxed{\text{F/W}}$ for at least half a second to leave menu mode.

8. Press $\boxed{\text{F/W}}$ $\boxed{3}$ to select power level. Scroll to the level you want (HIGH 5 W/MID 2 W/LOW 0.5 W).

9. Press and release $\boxed{\text{F/W}}$ to set power level.

10. Hold $\boxed{\text{F/W}}$ for at least half a second to start write.

11. Scroll to desired memory.

12. Press and release $\boxed{\text{F/W}}$ to write.

Lock/unlock radio

Press $\boxed{\text{F/W}}$ $\boxed{6}$ to lock or unlock.

Check repeater input frequency

Press and release [HM/RV] to switch between reverse and regular modes.

Change power in the field

Press and release [F/W] [3] (HIGH 5 W/MID 2 W/LOW 0.5 W). Press and release [F/W] to set.

Adjust volume

Rotate the center power/volume knob to adjust volume.

Adjust squelch

Rotate the outer ring of the selection knob to adjust RF squelch.

Weird Modes

If someone says you have Wires mode turned on

You are transmitting a DTMF tone that interrupts your first few syllables. Atom symbol appears in upper right of screen. Press and release the [0] key once. Atom symbol should go away.

Radio shows OUTRNG

This means you have ARTS mode enabled. Press and hold [4] (CODE) for two seconds to disable.

DTMF doesn't send memory

By default, the radio starts up in DTMF Memory mode. Pressing a number key while you hold [PTT] will send the number stored in that memory. Usually these memories are empty so nothing will be sent (although confusingly, [*] and [#] always transmit). To have the radio play the appropriate touchtone instead, press [F/W] [9] (DTMF). The radio will show CODE. At that point, pushing a number while [PTT] is held will send the touch tone. Press [F/W] [9] again to switch back to memory mode (MEM), if desired.

Useful Information

The time it takes for the [F/W] button to be registered as a "Hold" (as opposed to a "Press") is user-adjustable from 0.3 to 1.5 seconds. See Menu item 36 for details.

Factory reset

To reset, hold down $\boxed{\text{Moni}}$ and turn the radio on. Scroll to RSTALL (reset everything). Press $\boxed{\text{F/W}}$ to reset.

Memory reset

To reset, hold down $\boxed{\text{Moni}}$ and turn the radio on. Scroll to MEMRST (reset memories). Press $\boxed{\text{F/W}}$ to reset.

Memory bank reset

To reset, hold down $\boxed{\text{Moni}}$ and turn the radio on. Scroll to MB RST (reset memory banks). Press $\boxed{\text{F/W}}$ to reset.

Settings reset

To reset, hold down $\boxed{\text{Moni}}$ and turn the radio on. Scroll to SETRST (reset settings). Press $\boxed{\text{F/W}}$ to reset.

Radio Layout

Power/Volume

Emergency

PTT

Moni

F

P2
P1
P3
P4

Up

Down

*

#

Specs

Receivers Single receiver with priority channel option
Receives 65–108 MHz WFM, 136–174 MHz FM, 400–480 MHz FM
Transmits 144–148 MHz @ 5 W FM, 430–450 MHz @ 5 W FM
Antenna connector SMA **M** on radio; needs SMA **F** antenna
Modes FM
Memory Channels 200
Power No power input on radio, nominally 7.4 V DC
Model year 2017

Standard Tasks

Program frequency in the field

1. Enter VFO mode by pressing [*] if you're not already there. (Display shows M if you're in memory mode.)

2. Enter the frequency on the keypad (144390 for 144.390 MHz).

3. Hold [F] to enter menu mode.

4. To set the tone, use [Up] and [Down] to scroll to item 8, CTCSS. Press [F] to select TX if desired, then [Up] / [Down] to adjust. Press [F] to set. Similarly, use [Up] / [Down] to select RX if desired, then [Up] / [Down] to adjust. Press [F] to set.

5. To set shift, use [Up] and [Down] to scroll to item 24, Repeater. Press [F] to select Mode. Use [Up] / [Down] to switch between Simplex, +REP (positive offset) and -REP (negative offset). Press [F] to set.

6. If you need to adjust repeater offset frequency, use [Down] to select Shift. Press [F] to change, and use [Up] / [Down] to adjust. Press [F] to set when done.

7. To set tone type, use [Up] / [Down] to scroll to item 29, SQL Type. Press [F] to adjust. Use [Up] / [Down] to select from R-TONE (receive squelch), T-TONE (transmit tone), TSQL (transmit tone and receive tone squelch), REV TN (squelch if tone received), DCS (digital coded squelch), PAGER (unsquelch if two CTCSS tones received) , and OFF (no tone). Press [F] to set.

8. To adjust power, use [Up] / [Down] to go to item 32, TX PWR. Press [F] to adjust. Use [Up] / [Down] to choose from HI(5 W), MID(2.5 W) or LOW(0.5 W). Press [F] to set.

9. Hold [F] to exit the menu.

10. Hold [*] to start writing to memory.

11. Use [Up] / [Down] to select the memory.

12. hold [*] to complete writing to memory.

13. Press [*] to switch to memory mode if you're not already there.

14. Use [Up] / [Down] to scroll to the desired memory.

Lock/unlock radio

Hold [6] to lock or unlock.

Check repeater input frequency

Press [F][P4] to enter or exit reverse mode. Shift will blink when you're in reverse mode.

Change power in the field

To adjust power, hold [F] to enter the menu. Use [Up] / [Down] to go to item 32, TX PWR. Press [F] to adjust. Use [Up] / [Down] to choose from HI(5 W), MID(2.5 W) or LOW(0.5 W). Press [F] to set. Hold [F] to exit the menu.

Adjust volume

Rotate the knob to adjust volume.

Adjust squelch

Hold [F] to enter the menu. Scroll to item 26, RF SQL. Press [F] to adjust. Use [Up] / [Down] to go from OFF, S-1 through S-8, or S-FULL. Press [F] to set. Hold [F] to exit the menu.

Weird Modes

Radio is beeping continually

You are in emergency mode. Turn the radio off and then on. If you hold [Emergency] for three seconds, you will enter emergency mode. This causes a continual beep and wil transmit on the home frequency.

Can't enter VFO mode

This radio has a memory-only mode. To exit, turn the radio off. Then hold down [Moni] and [PTT] (does not transmit) while turning the radio on. Use [Down] to select F5:MEM-ONLY. Press [F] to toggle.

Can't enter VHF or UHF frequencies

This radio has an option to use VHF only or UHF only. To adjust it, turn the radio off. Then hold down [Moni] and [PTT] (does not transmit) while turning the radio on. Use [Down] to select F6:VHF-ONLY or F7:UHF-ONLY. Press [F] to toggle.

Radio shows OUT.RNG or IN.RNG

This means you have ARTS mode enabled. Hold [2] to exit.

Radio shows - - - - on power on

The password has been enabled. Enter the password (four digits) on the keypad. If you have forgotten the password, you can disable with a full reset (see below).

Useful Information

This radio has four programmable buttons which remember the radio state. Rather than programming a memory, once you've set up the frequency, hold [P1], [P2], [P3] or [P4] to program the button. Then later you can press [P1], [P2], [P3] or [P4] to recall the state (including frequency / tone info).

Factory reset

To reset, hold down [Moni] and [PTT] (does not transmit) while turning the radio on. Scroll to F4 ALL RESET (full reset—reset everything). Press [F] to reset.

Memory reset

To reset, hold down [Moni] and [PTT] (does not transmit) while turning the radio on. Scroll to F2 MEM RESET (reset memories). Press [F] to reset.

Memory bank reset

To reset, hold down [Moni] and [PTT] (does not transmit) while turning the radio on. Scroll to F3 BANK RESET (reset memory banks). Press [F] to reset.

Settings reset

To reset, hold down [Moni] and [PTT] (does not transmit) while turning the radio on. Scroll to F1 SET RESET (reset settings). Press [F] to reset.

Yaesu FT-70D

Radio Layout

Knob

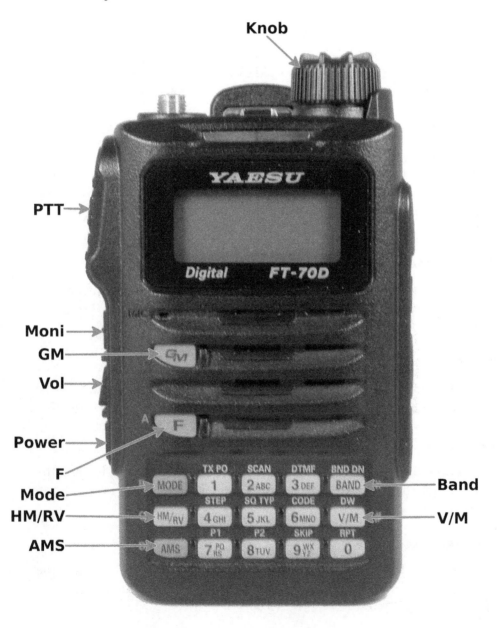

PTT

Moni

GM

Vol

Power

F

Mode

HM/RV

AMS

Band

V/M

Specs

Receivers Single receiver with priority channel option
Receives 108–136.995 MHz AM, 137–579.995 MHz FM
Transmits 144–147.995 MHz @ 5 W FM/C4FM, 430–449.995 MHz @ 5 W
 FM/C4FM
Antenna connector SMA F on radio; needs SMA M antenna
Modes FM, C4FM Fusion
Memory Channels 999
Power 6.0–16.0 V DC, nominally 7.4 V DC, EIAJ-02 barrel style, 4mm OD,
 1.7mm ID plug, center positive
Model year 2017

Standard Tasks

Program frequency in the field

 1. Press [V/M] to go to VFO mode if you aren't already there.

 2. Press [Band] repeatedly to cycle through bands and select the appro-
    priate one.

 3. Press [Mode] to set the mode (FM for analog FM mode, DN for Fusion
    digital mode).

 4. Use the number keys to enter frequency (144390 for 144.390 MHz).

 5. To change step size, press [F] [4]. Use the knob to adjust the step
    size. Press [PTT] (does not transmit) to set.

 6. To switch repeater shift, press [F] [0] to cycle through RPT- (negative
    offset), RPT+ (positive offset) and SIMP (simplex).

 7. If you need to set the repeater offset frequency, hold [F] for one sec-
    ond. Then use the knob to scroll to menu 46 RPT.FRQ. Press [F] to
    change, use the knob to adjust, then press PTT to set.

 8. To set tone mode, press [F] [5] repeatedly to cycle through OFF (no
    tone), TONE (send tone on transmit), TSQL (tone on transmit and re-
    ceive), DCS (digital coded squelch), RV TN (reverse tone—squelch if
    tone received), PR FRQ ("no-communication squelch" based on a tone),
    PAGER (play bell if two CTCSS tones received). Press [PTT] (does not
    transmit) to set.

 9. To set the tone frequency (after setting the tone mode), press [F] 6.
    Use the knob to adjust, and press [PTT] (does not transmit) to set.

10. Press [F] [1] repeatedly to select power level. Options are (HIGH 5
    W/MID 2 W/LOW 0.5 W). Press [PTT] (does not transmit) to set.

11. Hold [V/M] for at least one second to go to memory mode.

12. Use the knob to select the memory to write.

13. Press [V/M] to save the frequency.

14. You are presented with a second entry field. You can use this to set the name with the knob and [Band] keys.

15. Hold [V/M] to write.

16. Press [V/M] to go to memory mode.

17. Use the knob to scroll to the memory you want.

Lock/unlock radio

Presss [Power] to lock or unlock. Menu 30 LOCK lets you set what gets locked.

Check repeater input frequency

Press and release [HM/RV] to switch between reverse and regular modes.

Change power in the field

Press [F] [1] repeatedly to select power level. Options are HIGH (5 W) / MID (2 W) / LOW (0.5 W). Press [PTT] (does not transmit) to set.

Adjust volume

Hold [Vol] down and rotate the knob. Release [Vol].

Adjust squelch

Press [F] [Moni]. Use the knob to adjust squelch. Press [PTT] (does not transmit) to set.

Weird Modes

Radio shows GROUP and DG-ID

The Group Monitor function is on. Press [GM] to turn it off again.

Radio doesn't transmit

This radio has a PTT lock. Unlock the radio to transmit.
    You can change the lock setting by holding [F] to enter set mode, then scrolling to memory 30 LOCK. Press [F] to change, and use the knob to select the lock type you want.

Useful Information

Hold [Power] for about two seconds to turn off or on.
    Additional ways to send tones can be unlocked with menu 54 SQL.EXP.

Factory reset

Start by turning off the radio. To reset everything, hold [Mode], [HM/RV] and [AMS] while turning the receiver on. Then press [F] to reset.

Settings reset

Start by turning off the radio. To reset to default settings but save memories, hold down [Mode] and [V/M] while turning on. Then press [F] to reset.

Radio Layout

Specs

Receivers Single receiver
Receives 137–174 MHz FM
Transmits 144–148 MHz @ 5 W FM
Antenna connector SMA F on radio; needs SMA M antenna
Modes FM
Memory Channels 199
Power 13.8 V DC, EIAJ-02 barrel style, 4mm OD, 1.7mm ID plug, center
    positive
Model year 2000

Standard Tasks

Program frequency in the field

1. Press ⎡VFO⎤ to go to VFO mode if you aren't already there.

2. Enter frequency (144390 for 144.390 MHz).

3. Press ⎡F⎤⎡0⎤ to enter menu.

4. Scroll to Menu 26:TN SET or Menu 27:DCS SET. Press ⎡F⎤ to adjust value. Use the scroll knob for adjustment, and press ⎡F⎤ when done.

5. Scroll to Menu 25:SQL TYP. to adjust squelch type. Press ⎡F⎤ to adjust. There are four settings: TN ENC to send tone on transmit, TN SQL to send tone on transmit and require tone on receive, DCS and OFF. Use the scroll knob for adjustment, and press ⎡F⎤ when done.

6. Scroll to Menu 3:RPT to set repeater shift (SIMP/-RPT/+RPT). Press ⎡F⎤ to adjust. Use the scroll knob for adjustment, and press ⎡F⎤ when done.

7. Press ⎡PTT⎤ to leave menu mode (does not transmit).

8. Press ⎡F⎤⎡3⎤ to select power level. Scroll to the level you want (HIGH 5 W/MID 2 W/LOW 0.5 W).

9. Press and release ⎡F⎤ to set power level.

10. Hold ⎡F⎤ for at least one second to start write.

11. Scroll to desired memory within five seconds.

12. Press and release ⎡F⎤ to write.

13. Press ⎡MR⎤ to get to memory mode; scroll to the memory you just wrote.

Lock/unlock radio

Press ⎡F⎤⎡6⎤ to lock or unlock.

Check repeater input frequency

Press and release REV to switch between reverse and regular modes.

Change power in the field

Press F 3 to select power level. Scroll to the level you want (HIGH 5 W/MID 2 W/LOW 0.5 W). Press F to set.

Adjust volume

Rotate the center power/volume knob to adjust volume.

Adjust squelch

Rotate the outer ring of the selection knob to adjust squelch.

Weird Modes

Radio shows OUT RNG

This means you have ARTS mode enabled. Press F to exit ARTS mode.

Useful Information

F 0 Menu 37:BATT F will show battery info.

Factory reset

To reset the radio to factory settings, hold down Lamp and PTT while turning the radio on (does not transmit). Scroll to ALL.RST (reset settings and memories). Press F to reset.

Settings reset

To reset the radio settings only but keep the memories, hold down Lamp and PTT while turning the radio on (does not transmit). Scroll to SET.RST (reset settings). Press F to reset.

Yaesu VX-170

Radio Layout

Power/Volume

Ring

PTT

Moni

REV

F

MR

VFO

Specs

Receivers Single receiver with priority channel option
Receives 137–174 MHz FM
Transmits 144–148 MHz @ 5 W FM
Antenna connector SMA F on radio; needs SMA M antenna
Modes FM
Memory Channels 200
Power 6.0–16.0 V DC, EIAJ-02 barrel style, 4mm OD, 1.7mm ID plug, center positive
Model year 2005

Standard Tasks

Program frequency in the field

1. Press VFO to go to VFO mode if you aren't already there. (Display will show -A- or -b- in VFO mode, and a memory number in memory mode.)

2. Enter frequency (144390 for 144.390 MHz).

3. Press F 0 to enter menu.

4. Scroll to Menu 44:SQL.TYP to adjust squelch type. Press F to adjust. There are six settings: TONE to send tone on transmit, TSQL to send tone on transmit and require tone on receive, REV TN to mute when tone received, DCS digital coded squelch, ECS "enhanced paging," and OFF. Use the outer ring for adjustment, and press F when done.

5. Scroll to Menu 46:TN FRQ or Menu 13:DCS.COD. Press F to adjust value. Use the outer ring for adjustment, and press F when done.

6. Scroll to Menu 35:RPT.MOD to set repeater shift (RPT. + / RPT. - / RPT.OFF). Press F to adjust. Use the outer ring for adjustment, and press F when done.

7. If needed, scroll to Menu 41:SHIFT to set repeater offset. Press F to adjust. Use the outer ring for adjustment, and press F when done.

8. Press PTT to leave menu mode (does not transmit).

9. Press F 3 to select power level. Scroll to the level you want (HIGH 5 W/MID 2 W/LOW 0.5 W).

10. Press and release F to set power level.

11. Hold F for at least one second to start write.

12. Scroll to desired memory within five seconds.

13. Press and release F to write.

14. Press MR to go to memory mode; scroll to the memory you just wrote.

Lock/unlock radio

Press [F] [6] to lock or unlock.

Check repeater input frequency

Press and release [REV] to switch between reverse and regular modes.

Change power in the field

Press [F] [3] to select power level. Scroll to the level you want (HIGH 5 W/MID 2 W/LOW 0.5 W). Press [F] to set.

Adjust volume

Rotate the center power/volume knob to adjust volume.

Adjust squelch

Press [F] [Moni]. Rotate the outer ring to adjust squelch. Press [F] to save.

Weird Modes

Radio shows OUT.RNG

This means you have ARTS mode enabled. Press [F] to exit ARTS mode.

Can't enter VFO mode

This radio has a memory-only mode. To enable/disable it, hold down [Moni] while turning the radio on. Scroll to F5 M-ONLY and press [F].

Useful Information

[F] [0] Menu 12:DC VLT [F] will show battery info.

Factory reset

Hold down [Moni] while turning the radio on. Use the outer ring to select F4 ALLRST (reset everything). Press [F] to reset.

Memory reset

Hold down [Moni] while turning the radio on. Use the outer ring to select F2 MEMRST (clear memories). Press [F] to reset.

Memory bank reset

Hold down [Moni] while turning the radio on. Use the outer ring to select F3 MB RST (clear memory banks). Press [F] to reset.

Settings reset

Hold down Moni while turning the radio on. Use the outer ring to select
F1 SETRST (reset settings). Press F to reset.

Yaesu VX-1R

Radio Layout

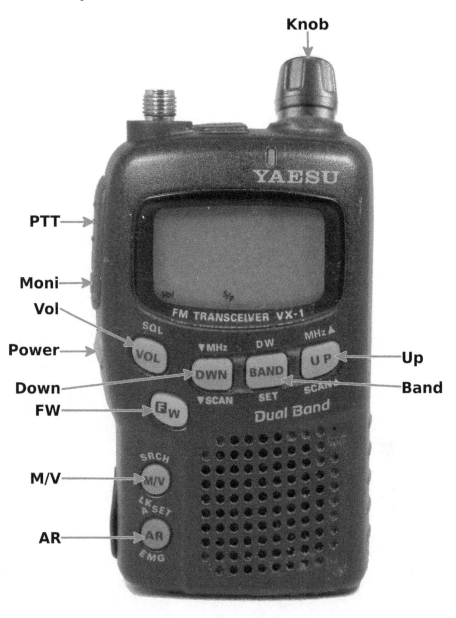

Specs

Receivers Single receiver, dual watch (first to break squelch wins)
Receives 0.5–1.7 MHz AM, 76–108 MHz FM, 108–137 MHz AM, 137–999 MHz FM less cellular
Transmits 144–148 MHz @ 0.5 W FM, 430–450 MHz FM @ 0.5 W
Antenna connector SMA F on radio; needs SMA M antenna
Modes FM
Memory Channels 52 in group 1, 142 in group 2
Power 3.2–7.0 V DC, EIAJ-01 barrel style, 2.35mm OD, 0.70mm ID plug, center positive
Model year 1997

Standard Tasks

Program frequency in the field

1. Press M/V to go to VFO mode if you aren't already there. (Memory mode shows a channel number above the frequency; VFO mode does not.)

2. Press Band to select either V-HAM (2 m) or U-HAM (70 cm).

3. Enter frequency using the knob. Press FW to change in 1 MHz steps; press FW again to go back to normal.

4. Hold Band to enter menu.

5. Press Up or Down to get to menu 25:T SET or Menu 27:DCS. Use knob to adjust value.

6. Press Down to scroll to menu 24:SQL TYP. to adjust squelch/tone type. Use the knob to select T (tone), T SQ (tone squelch) or DCS.

7. Press Down to scroll to menu 6:SHIFT to set repeater offset. Use the knob to adjust (affects VFO shift as well).

8. The radio has automatic repeater shift. If you want to disable this, press Down to scroll to menu 4: ARS and use the knob to select ARS OFF.

9. To set repeater shift manually after turning ARS off, press Up to scroll to 5: RPTR. Use the knob to select +, - or no shift.

10. To set transmit power, use Up and Down to scroll to 1:TX PWR. Use the knob to adjust the value to HIGH (0.5w) or LOW (50 mW).

11. Press Band to leave menu mode (does not transmit).

12. Hold FW to enter memory write mode.

13. Use the knob to select the memory to write into. Press FW to write.

14. Press M/V to get to memory mode; scroll to the memory you just wrote.

Lock/unlock radio

Hold M/V for one second to lock or unlock.

Check repeater input frequency

Press FW and hold Moni. Releasing Moni switches back to normal mode.

Change power in the field

Hold Band for one second to enter menu mode, then use Up and Down to scroll to 1:TX PWR. Use the knob to adjust the value to HIGH (0.5w) or LOW (50 mW). Press Band to exit.

Adjust volume

Press Vol, then within two seconds use knob to set volume (1–31 or MUTE). Volume mode automatically exits after two seconds.

Adjust squelch

The radio is in "Auto-squelch" mode by default. To adjust, press FW then Vol. Then within two seconds use the knob to adjust from SQL AUT (auto), SQL OPEN (open) and squelch values 1–10. Squelch mode automatically exits after two seconds.

Weird Modes

Knob doesn't change memory

This radio's knob can either adjust volume / squelch or change memory. By default it selects memory, but you can use 29:DIAL M to switch it to VOL/SQL or DIAL.

Radio is beeping

The radio has an Emergency Mode that you enter by holding AR for two seconds. Turn the radio off then on again to exit.

Radio shows OUTRNG or IN RNG

This means you have ARTS mode enabled. Press AR to exit ARTS mode.

Useful Information

Menu 14: LOCK controls what the radio can do when locked. Options are KEY key lock, DIAL knob lock, D + K (knob and key lock), PTT (PTT lock—no transmit), K + P (key + PTT lock), D + P (knob + PTT lock) and ALL (everything locked).

Memory group 2 is intended for simplex frequencies. It can store repeater shift and tone mode but not tone frequency.

The Moni button looks like it's part of the PTT button. It's really a separate button.

Factory reset

To reset the radio entirely (including memories), hold down M/V and AR while turning the radio on, then release all buttons. Display will show INI? F. Press FW to reset. Display will show INITIAL.

Settings reset

To reset the radio to default settings (does not affect memories), hold down FW and Vol while turning the radio on, then release all buttons. Display will show SETINN.

Yaesu VX-6R

Radio Layout

Specs

Receivers Single receiver
Receives 0.5–998.900 MHz
Transmits 144–148 MHz @ 5 W FM, 222–225 MHz @ 1.5 W FM, 430–450 MHz @ 5 W FM
Antenna connector SMA F on radio; needs SMA M antenna
Modes FM
Memory Channels 1000
Power 5–16 V DC, EIAJ-02 barrel style, 4mm OD, 1.7mm ID plug, center positive
Model year 2005

Standard Tasks

Program frequency in the field

1. Press V/M to go to VFO mode if you aren't already there.

2. Enter frequency (144390 for 144.390 MHz).

3. Press F/W then 0 to enter menu mode.

4. To set squelch type, rotate ring to 60: SQL TYPE. Press 0 then use ring to scroll through OFF, TONE, TSQL and DCS. Press 0 to set.

5. To set tone frequency, rotate ring to 66: TN FREQ. Press 0 to adjust, then use ring to select tone. Press 0 to set.

6. To set repeater shift direction, scroll with ring to 51: RPT. Press 0 to cycle through -RPT, +RPT and SIMP. Press 0 to set.

7. To set repeater offset, scroll with ring to 56: SHIFT. Press 0 to change, then use ring to select value (0.60 is 600 kHz). Press 0 to set.

8. Press PTT to leave menu mode (does not transmit).

9. Press F/W then Wires to begin setting power level.

10. Press Wires to select among LOW1 (0.3 W, 0.2 W on 220), LOW2 (1 W, 0.5 W on 220), LOW3 (2.5 W, 1.0 W on 220) and HIGH (5 W, 1.5 W on 220).

11. Wait three seconds to leave power setting mode.

12. Hold F/W for at least one second. F will flash.

13. Use ring to scroll to desired memory (flashing channel number means empty). Press and release F/W to write.

14. Press V/M to get to Memory mode.

Lock/unlock radio

Press and hold [Wires] for at least two seconds to lock/unlock. While unlocking, the radio will play a tone that you can ignore. Note that the lock style can be adjusted.

Check repeater input frequency

Press and release [HM/RV] to switch between reverse and regular modes. Shift direction will flash when in reverse mode.

Change power in the field

Press [F/W] then [Wires] to begin setting power level. Then press [Wires] to select among LOW1 (0.3 W, 0.2 W on 220), LOW2 (1 W, 0.5 W on 220), LOW3 (2.5 W, 1.0 W on 220) and HIGH (5 W, 1.5 W on 220). Wait three seconds to leave power setting mode.

Adjust volume

Rotate the knob to adjust volume.

Adjust squelch

Press [F/W] then [0] to enter set mode. Use ring to scroll to 59: SQL. Press [0] to change. Rotate ring to choose from LVL 0 (no squelch) to LVL 15. Press [0] to set, then [PTT] to leave menu mode.

Weird Modes

If someone says you have Wires mode turned on

You are transmitting a DTMF tone that interrupts your first few syllables. Press and release the [Wires] key once. Atom symbol on the right of the display will disappear.

Radio shows OUT RANGE

This means you have ARTS mode enabled. Press [4] (ARTS) to disable.

Useful Information

Menu 35: LOCK lets you select from KEY, DIAL, K+D, PTT, P+K, P+D, and ALL for the lock mode.

Factory reset

To reset everything, hold down [0], [Mode] and [V/M] while turning the radio on, then press [F/W] to reset.

Settings reset

To reset all menu settings but leave memories intact, hold down [Mode] and [V/M] while turning the radio on, then press [F/W] to reset.

Yaesu VX-7R

Radio Layout

Knob/Volume

Ring

PTT

Power

Main

Sub

Mon F

HM/RV

Wires

V/M

Specs

Receivers Two independent receivers, simultaneous receive
Receives 0.5–30 MHz, 50–999 MHz on main band, 50–54 MHz, 140–174
    MHz, 420–470 MHz on sub band
Transmits 50–54 MHz @ 5 W FM, 144–148 MHz @ 5 W FM, 420–470 MHz
    @ 5 W FM, 222–225 MHz @ 0.3 W FM, 50–54 MHz @ 1 W AM
Antenna connector SMA F on radio; needs SMA M antenna
Modes FM, AM (main VFO only)
Memory Channels 450
Power 10–16 V DC, EIAJ-02 barrel style, 4mm OD, 1.7mm ID plug, center
    positive
Model year 2002

Standard Tasks

Program frequency in the field

1. Press V/M to go to VFO mode if you aren't already there.

2. Enter frequency (144390 for 144.390 MHz).

3. Press Mon F then 0. Scroll to Menu TSQ/DCS/DTMF:1 SQL TYPE.
   Press Sub to scroll through OFF, DCS, TONE SQL and TONE.

4. Rotate outer ring clockwise to TSQ/DCS/DTMF:2 TONE SET. Press Band
   to adjust, then use Main and Sub to scroll up and down to select
   tone. Press Band when done.

5. Press PTT to leave menu mode (does not transmit).

6. Press Mon F then Wires to begin setting power level.

7. Press Wires to select among L1 (0.05 W), L2 (1 W), L3 (2.5 W) and no
   indication (5 W).

8. Press Mon F to leave power setting mode.

9. Hold Mon F for at least half a second.

10. Use outer ring to scroll to desired memory (* means empty). Press
    and release Mon F to write.

11. By default, this radio has automatic repeater shift (ARS). Press Mon F
    then 0 then scroll with outer ring to Basic Setup:5. Press Sub to
    switch it ON if it's not.

12. If you need to enter a different transmit frequency (odd split âĂŞ not
    the one from ARS), enter the transmit frequency using the keypad.
    Then Press Mon F for half a second. Within five seconds, rotate
    the outer ring to the same memory you programmed the receive fre-
    quency on in step 10. Press and hold PTT and while holding PTT
    down press Mon F. This does not transmit. The repeater transmit
    frequency is stored.

13. Press V/M to get to Memory mode.

Lock/unlock radio

Press and hold Wires for at least two seconds. Note that the Lock style can be adjusted.

Check repeater input frequency

Press and release HM/RV to switch between reverse and regular modes.

Change power in the field

Press and release Mon F. Press and release Wires key repeatedly to cycle through L1 (0.05 W), L2 (1 W), L3 (2.5 W) and no indication (5 W). Press and release Mon F to set.

Adjust volume

Rotate the inner knob to adjust volume.

Adjust squelch

Rotate the outer ring to adjust squelch.

Weird Modes

If someone says you have Wires mode turned on

You are transmitting a DTMF tone that interrupts your first few syllables. Press and release the Wires key once. Radio/PC/Globe symbol will disappear.

Radio shows OUT RANGE

This means you have ARTS mode enabled. Press Mon F then 4 (ARTS) to disable.

Useful Information

The order of menu sections is Basic Setup (1–14), Display Setup (1–8), TSQ/DCS/DTMF (1–8), Scan Modes (1–7), Measurement (1–7), Save Modes (1–6), ARTS (1–3), Misc Setup (1–20).

This radio comes in two colors: black and silver.

Factory reset

To reset everything, hold down 4, Band and V/M while turning the radio on.

Settings reset

To reset all settings but leave memories intact, hold down Band and V/M while turning the radio on.

Yaesu VX-8DR

Radio Layout

Specs

Receivers Two independent receivers, simultaneous receive
Receives 510–1790 kHz AM on VFO A, 0.5–30 MHz AM on VFO A, 30–
  76MHz FM on VFO A, 108–999.9 MHz FM on VFO A (cell blocked),
  108–580 MHz FM on VFO B
Transmits 50–54 MHz @ 5 W FM, 144–148 MHz @ 5 W FM, 440–450 MHz
  @5 W FM, 222–225 MHz @1.5 W FM, 50–54 MHz @ 1 W AM
Antenna connector SMA F on radio; needs SMA M antenna
Modes FM, AM (VFO A only)
Memory Channels 900
Power 11–14 V DC, EIAJ-02 barrel style, 4mm OD, 1.7mm ID plug, center
  positive
Model year 2009

Standard Tasks

Program frequency in the field

1. Press $\boxed{\text{V/M}}$ to go to VFO mode if you aren't already there.

2. Enter frequency (144390 for 144.390 MHz).

3. Hold $\boxed{\text{Menu}}$ for at least half a second.

4. Scroll to Menu 99:Set Tone Frequency then press $\boxed{\text{Menu}}$ to adjust.
   Use the scroll knob for adjustment, and press $\boxed{\text{Menu}}$ when done.

5. Scroll to Menu 95:Set SQL Type (options are TONE, TONE SQL, DCS,
   REV TONE (mutes when tone received), PR FREQ (mutes when PR frequency tone is received), PAGER (pager tone received), MESSAGE (APRS
   message recieved) or OFF) then press $\boxed{\text{Menu}}$ to adjust. Use the scroll
   knob for adjustment, and press $\boxed{\text{Menu}}$ when done.

6. Scroll to Menu 75:Set repeater shift (-RPT/+RPT/SIMPLEX) then
   press $\boxed{\text{Menu}}$ to adjust. Use the scroll knob for adjustment, and press
   $\boxed{\text{Menu}}$ when done.

7. Hold $\boxed{\text{Menu}}$ for at least half a second.

8. Press and release $\boxed{\text{F/W}}$.

9. Press and release $\boxed{\text{Wires}}$ key repeatedly to select power level. Options
   are L1 (50 mW), L2 (1.5 W), L3 (2.5 W) and HI (5 W).

10. Press and release $\boxed{\text{F/W}}$ to set power level.

11. Hold $\boxed{\text{F/W}}$ for at least half a second to start write.

12. Scroll to desired memory.

13. Press and release $\boxed{\text{F/W}}$ to write.

14. Press $\boxed{\text{V/M}}$ to get to Memory mode.

Lock/unlock radio

Press and release [Power] button for less than half a second to lock and unlock.

Check repeater input frequency

Press and release [HM/RV] to switch between reverse and regular modes.

Change power in the field

Press and release [F/W]. Press and release [Wires] key repeatedly to select power level. Options are L1 (50 mW), L2 (1.5 W), L3 (2.5 W) and HI (5 W). Press and release [F/W] to set.

Adjust volume

Hold down [Vol] and rotate the scroll knob to adjust. Each VFO has its own volume.

Adjust squelch

Hold Menu for at least half a second. Scroll to Menu 92: Set SQL Level then press Menu to adjust. Use the scroll knob for adjustment, and press Menu when done. Hold Menu for at least half a second

Weird Modes

If someone says you have Wires mode turned on

You are transmitting a DTMF tone that interrupts your first few syllables. Press and release the [Wires] key once. Atom symbol will disappear.

Radio shows OUTRNG

This means you have ARTS mode enabled. Press and release [4] (ARTS) to disable.

Radio volume control stuck or reversed

The VX-8 has a feature for mobile use that will reverse how the volume adjustment behaves. If this feature is enabled, the knob and arrows will adjust volume but not frequency. Holding down the volume control with this feature enabled will provide "normal" behavior. To get in or out of this mode, press and release the [F/W] key, then press and release the [Vol] key.

Useful Information

The radio can have GPS installed on radio body (mic connector) or on speaker mic. If you don't need it, you can remove it to save battery (needs a screwdriver).

Hold down the button for VFO A or VFO B to switch between monitoring two frequencies and monitoring one frequency.

Factory reset

To reset everything, hold down ⎡Band⎤, ⎡HM/RV⎤ and ⎡Wires⎤ while turning the radio on, then press ⎡F/W⎤.

Settings reset

To reset most settings but leave memories intact, hold down ⎡Band⎤ and ⎡V/M⎤ while turning the radio on, then press ⎡F/W⎤.

Zastone ZT-V8

Radio Layout

Specs

Receivers Single receiver, dual watch (first to break squelch wins)
Receives 136–174 MHz and 400–480 MHz FM
Transmits 136–174 MHz @ 4 W FM and 400–480 MHz @ 4 W FM
Antenna connector SMA **M** on radio; needs SMA **F** antenna
Modes FM
Memory Channels 128
Power No DC input on radio
Model year 2012

Standard Tasks

Program frequency in the field

1. Decide which memory to use. You can't overwrite—you will have to delete first if that memory has data in it. Press [Menu] [2] [8] [Menu] (DEL-CH) *XXX* where *XXX* is the channel (001–128). Then press [Menu].

2. Press [VFO/MR] to go to VFO mode if you aren't there already.

3. Enter frequency (144390 for 144.390 MHz). You may need to press [Band] to switch bands depending on your frequency.

4. Press [Menu] [1] [3] [Menu] (T-CTCS) then scroll to the correct CTCSS tone frequency (or off), press [Menu] [Exit]. Scroll using the [Up] and [Down] arrow keys.

5. Press [Menu] [2] [5] [Menu] (SFT-D) and scroll to the correct repeater shift, press [Menu] [Exit].

6. Press [Menu] [2] [6] [Menu] (OFFSET) and enter or scroll to the correct repeater offset, then press [Menu] [Exit].

7. Press [Menu] [2] [Menu] (TXP) and scroll to correct power level, then press [Menu] [Exit].

8. Press [Menu] [2] [7] [Menu] (MEM-CH) and enter channel to write *XXX* (001–128) then press [Menu] [Exit].

9. Press [VFO/MR] to enter channel (memory) mode.

10. Scroll to the channel you just wrote.

Lock/unlock radio

Hold [#] for three seconds to lock/unlock.

Check repeater input frequency

There is no way to listen to the repeater input frequency except to program a separate memory for it.

Change power in the field

Press [Menu] [2] [Menu] (TXP) and scroll to correct power level (LOW is 1 W, HIGH is 4 W), press [Menu] [Exit]

Adjust volume

Rotate power/volume knob to adjust volume.

Adjust squelch

Press [Menu] [0] [Menu] (SQL) then scroll to the correct squelch level, press [Menu] [Exit].

Weird Modes

Can't leave channel (memory) mode

Some distributors ship these radios with VFO mode turned off. If that's the case, you will need to modify the programming with a computer / radio programming cable to turn VFO mode on. This cannot be changed from the front panel.

Can't set offset / direction

This radio can display frequencies instead of channel names. If you have that enabled, it's difficult to tell if you're in channel (memory) mode or frequency (VFO) mode. One sure way is to try to program an offset or shift. If it doesn't "stick" after programming, that's likely because you're in channel mode. Switch to frequency mode in order to program.

Useful Information

This appears to be a re-badged Baofeng UV-5R with modified firmware.

Hold down [3] while turning the radio on to see the firmware version.

If you wait too long after pressing [Menu], it will time out. You need to press quickly.

No reset procedure

This radio cannot be reset from the front panel.

⚠ **WARNING:** In at least some configurations, the Zastone ZT-V8 may permit you to transmit on business or public safety frequencies. Make sure you are in-band when transmitting.

Acknowledgments

It's very easy to end up owning too many handhelds. Thanks to the following people, who very graciously allowed me to use their handhelds and let me mess up their programming in the process of writing this book:

Michael Drapkin, WB2SEF, lent me his Icom IC-P2AT and Icom IC-2AT. (Or maybe one of them was his wife Jennifer N7ZID's radio that Michael lent me.)

Tsuyoshi Nagano, WH7JJ, lent me his Kenwood TH-D72A.

Rich Langevin, KB7YEB, lent me his Kenwood TH-D7 and Yaesu FT-50.

Ed Karsten, K9EDK, gave me his Icom IC-W32A.

Jim Pierce, N7QVW, lent me his Icom IC-T70A and Yaesu VX-170.

Dennis Bietry, KE7EJF, let me play with his Harris Unity XG-100P.

Julie Jennings, KI6UCJ, lent me her Kenwood TH-F6a.

Walt Reinert, NJ8G, lent me his Yaesu VX-6R.

Phil Arnold, KJ6STS, lent me his Kenwood TH-78A.

Arnold Knack, KA7AOK, let me photograph his Icom ID-51A.

Thanks to the great people at Ham Radio Outlet in Phoenix, who let me photograph and play with the the Yaesu FT-70D, Yaesu FT-2D, Icom IC-V80 HD, Kenwood TH-K20A, Alinco DJ-500T, Kenwood TH-D74, Yaesu FT-25, Yaesu FT-65, Yaesu FT-270, Tera TR-590 and Wouxun KG-UV9D. In particular I'd like to thank Mike Spraker KB6VHF, Jim Skinnell K7EOG and Mike Baker K7DD. Support your local ham store!

Thanks to Jim Pierce, N7QVW, for suggesting the chapters on batteries and pre-programmable radios. Thanks to Andy Durbin, K3WYC, for the tip on Wouxun dealer mode. Thanks to Michal Drapkin, WB2SEF, for suggesting the glossary chapter and for his careful review.

Thanks also to everyone in the Maricopa County Emergency Communications Group, for providing a way for hams to learn about their radios in the field while helping others.

Thanks to Renée Targos for her help with editing. Any spelling errors you don't see are due to her.

Thanks to the creators of TEXstudio, LaTeX, the Memoir style, TikZ, TEX and GIMP for making the wonderful tools I used for this book.

Thanks to my father, Andy, for leaving the license manuals at my house.

And finally, thanks to my wife, Debbie, who likes it when I put antennas in the yard.

References

[1] Coaxial Power Connector. In *Wikipedia*. Retrieved from `https://en.wikipedia.org/wiki/Coaxial_power_connector`

[2] *ECR-5892D: Comparison of NiCd, NiMH, and Li-Ion Batteries*. Retrieved from `https://www.portal.state.pa.us/portal/server.pt/directory/technical_publications`

[3] EIAJ Connector. In *Wikipedia*. Retrieved from `https://en.wikipedia.org/wiki/EIAJ_connector`

[4] Emission Classification. In *Amateur Radio Wiki—The Online Encyclopedia for Hams*. Retrieved from `http://www.amateur-radio-wiki.net/index.php?title=Emission_Classification`

[5] *FCC ID Application Database*. `http://fccid.io`

[6] *Hampedia*. `http://hampedia.net`

[7] *Internet Archive*. `http://archive.org`

[8] *RigPix*. `http://www.rigpix.com`

[9] *RigReference.com*. `http://rigreference.com`

[10] *SARL Radio Age and Price Guide*. `http://www.sarl.org.za/Radios.asp`

[11] Understanding the Radioshack Adaptaplug System. In *RadioShack Guide to Understanding Power Conversion*. Retrieved from `http://support.radioshack.com/support_tutorials/batteries/pwrgde-2H.htm`

CPSIA information can be obtained
at www.ICGtesting.com
Printed in the USA
LVHW101326271118
598410LV00018B/183/P